SRA
Essentials for
Algebra

A Direct Instruction Approach

Workbook

Siegfried Engelmann
Bernadette Kelly
Owen Engelmann

McGraw Hill **SRA**

Columbus, OH

Front Cover Photo: © Lester Lefkowitz/Getty Images, Inc.

SRAonline.com

Send all inquiries to this address:
SRA/McGraw-Hill
4400 Easton Commons
Columbus, OH 43219

ISBN: 978-0-07-602193-2
MHID: 0-07-602193-9

9 QVS 13

The McGraw-Hill Companies

Table of CONTENTS

Table of CONTENTS

Lesson 1

Part 1 ▸ Cross out the problems you **cannot** work the way they are written. Work the problems you **can** work the way they are written.

> ① Copy the denominator.
> ② Add or subtract for the numerator.

a. $\dfrac{2}{7} + \dfrac{11}{7} =$

b. $\dfrac{12}{c} - \dfrac{10}{c} =$

c. $\dfrac{6}{m} + \dfrac{5}{2m} =$

d. $\dfrac{15}{14} + \dfrac{2}{14} =$

e. $\dfrac{10}{3} - \dfrac{5}{4} =$

f. $\dfrac{11}{15} - \dfrac{4}{6} =$

g. $\dfrac{5}{10} - \dfrac{4}{10} =$

h. $\dfrac{9}{9} + \dfrac{4}{7} =$

i. $\dfrac{30}{b} - \dfrac{20}{b} =$

Lesson 2

Part 1 Write the abbreviation for each unit name.

Unit Name	Abbreviation
miles	_____
centimeters	_____
yards	_____
meters	_____
feet	_____
kilometers	_____
inches	_____

Part 2 Cross out the problems you **cannot** work the way they are written. Work the problems you **can** work the way they are written.

a. $4 - \dfrac{3}{5} =$

b. $\dfrac{9}{12} - \dfrac{7}{12} =$

c. $\dfrac{20}{m} + \dfrac{7}{m} =$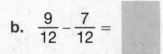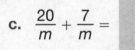

d. $\dfrac{11}{c} + \dfrac{11}{d} =$

e. $1 - \dfrac{5}{9} =$

f. $\dfrac{13}{35} + \dfrac{4}{35} =$

g. $\dfrac{3}{5} + \dfrac{5}{3} =$

h. $\dfrac{10}{2n} - \dfrac{9}{2n} =$

Lesson 3

Part 1 ▸ Cross out the problems you **cannot** work the way they are written. Work the problems you **can** work the way they are written.

a. $\dfrac{2}{3} \times \dfrac{1}{5} =$

d. $\dfrac{12}{9} \times \dfrac{2}{7} =$

g. $\dfrac{1}{20} \times \dfrac{23}{3} =$

b. $\dfrac{2}{3} + \dfrac{1}{5} =$

e. $\dfrac{10}{1} \times \dfrac{3}{8} =$

c. $\dfrac{8}{p} + \dfrac{17}{d} =$

f. $10 - \dfrac{3}{8} =$

Lesson 4

Part 1 ▸ Adding and Subtracting Decimal Values

◆ When you add or subtract decimal values, write those values with the decimal points lined up.

👎 a.
```
    6.7 5
    .1 8 2
   28.9
    1.0 0 4
+ 11 0.2 2
```

👍 b.
```
    6.75
    .182
   28.9
    1.004
+ 110.22
```

Lesson 7

Part 1 ▸ Multiply each whole number by the fractions $\dfrac{2}{2}, \dfrac{3}{3}, \dfrac{4}{4},$ and $\dfrac{5}{5}$. Complete each equation.

a. $5 = \dfrac{}{} = \dfrac{}{} = \dfrac{}{} = \dfrac{}{}$

b. $3 = \dfrac{}{} = \dfrac{}{} = \dfrac{}{} = \dfrac{}{}$

Lesson 8

$$\frac{4}{5}\left(\frac{7}{7}\right) = \frac{28}{35}$$

So: $\boxed{\dfrac{4}{5} = \dfrac{28}{35}}$

b. $\dfrac{3}{4} \times \dfrac{5}{5} =$ ▮

▮ = ▮

d. $\dfrac{8}{10} \times \dfrac{10}{10} =$ ▮

▮ = ▮

a. $\dfrac{5}{9}\left(\dfrac{2}{2}\right) =$ ▮

$\dfrac{5}{9} =$ ▮

c. $\dfrac{2}{7}\left(\dfrac{9}{9}\right) =$ ▮

▮ = ▮

e. $\dfrac{11}{5}\left(\dfrac{7}{7}\right) =$ ▮

▮ = ▮

Part 2 ➤ Multiply the whole number by the fractions $\frac{2}{2}$, $\frac{3}{3}$, $\frac{4}{4}$, and $\frac{5}{5}$. Complete the equation.

$$10 = \frac{}{} = \frac{}{} = \frac{}{} = \frac{}{}$$

Part 3 ➤ Fill in the missing numerators.

$$4 = \frac{}{7} = \frac{}{3} = \frac{}{6} = \frac{}{2}$$

Lesson 10

◆ Rewrite the whole number so you can add it to the fraction.

$$7 \frac{3}{5}$$

$$\frac{35}{5} + \frac{3}{5} = \boxed{\frac{38}{5}}$$

◆ **For each mixed number, write an addition equation below.**

a. $6 \frac{2}{3}$

_____

d. $3 \frac{4}{11}$

_____

b. $10 \frac{5}{6}$

_____

e. $5 \frac{4}{7}$

_____

c. $7 \frac{2}{9}$

_____

f. $8 \frac{2}{3}$

_____

Lesson 11

Part 1 Write an addition equation for each mixed number.

a. $3\frac{9}{10}$

d. $4\frac{5}{7}$

_____

_____

b. $25\frac{1}{2}$

e. $20\frac{5}{6}$

_____

_____

c. $8\frac{1}{3}$

f. $7\frac{3}{8}$

_____

_____

Lesson 12

Part 1 Percent

♦ A **percent** number is a **hundredths** number:

$$16\% = \frac{16}{100}$$

♦ You can rewrite any hundredths fraction as a percent:

$$\frac{45}{100} = 45\%$$

♦ **Complete the table.**

	Fraction	Percent
a.	$\frac{285}{100}$	
b.		16%
c.	$\frac{7}{100}$	
d.	$\frac{100}{100}$	
e.		325%

Part 2 Write the fraction you multiply $\frac{3}{4}$ by to get each fraction shown.

≠ means **not** equal

♦ **Change the sign in front of each fraction that does not equal $\frac{3}{4}$.**

$$\frac{3}{4} = \frac{15}{20} = \frac{6}{8} = \frac{30}{36} = \frac{12}{12} = \frac{21}{28} = \frac{18}{20}$$

Lesson 13

Part 1 Write the fraction you multiply $\frac{2}{10}$ by to get each fraction shown. Some fractions need the ≠ sign.

$$\frac{2}{10} = \frac{20}{80} = \frac{12}{60} = \frac{18}{90} = \frac{10}{40} = \frac{10}{50}$$

Part 2 Decimal Values for Hundredths

$$\frac{4}{100} = .04 \qquad \Big| \qquad \frac{536}{100} = 5.36$$

◆ Complete the table.

	Decimal	Fraction	Percent
	2.06	$\frac{206}{100}$	206%
a.		$\frac{186}{100}$	
b.		$\frac{80}{100}$	
c.		$\frac{7}{100}$	
d.		$\frac{15}{100}$	
e.		$\frac{258}{100}$	

Lesson 14

Part 1 ▶ Write the fraction you multiply $\frac{5}{3}$ by to get each fraction shown. Some fractions need the $\neq$ sign.

$$\frac{5}{3} \;=\; \frac{20}{12} \;=\; \frac{25}{18} \;=\; \frac{50}{30} \;=\; \frac{15}{15}$$

Part 2 ▶ Write equivalent decimal values for the percents your teacher says.

a. _____   b. _____   c. _____

Part 3 ▶ Complete the table.

	Decimal	Fraction	Percent
a.			142%
b.			9%
c.			28%
d.			518%
e.			3%
f.			70%

Lesson 15

Write the fraction you multiply $\frac{9}{5}$ by to get each fraction shown. Some fractions need the ≠ sign.

$$\frac{9}{5} = \frac{45}{20} = \frac{27}{20} = \frac{54}{30} = \frac{90}{50} = \frac{63}{40}$$

Part 2 Complete the table.

	Decimal	Fraction	Percent
a.		$\frac{100}{100}$	
b.			802%
c.	4.03		
d.			8%
e.		$\frac{90}{100}$	

Lesson 16

Part 1 Complete the division problem and the answer for each fraction. Box each answer.

a. $\dfrac{24}{4}$ b. $\dfrac{50}{10}$ c. $\dfrac{86}{2}$ d. $\dfrac{99}{9}$

Lesson 17

Part 1 Divisibility Rules

34	55	70

◆ Circle what each number is divisible by.
Some numbers are divisible by more than one value.

a. 260

divisible by:

10	5	2

c. 364

10	5	2

e. 32

10	5	2

b. 45

divisible by:

10	5	2

d. 10

10	5	2

f. 205

10	5	2

Lesson 19

♦ When you add, you can show the values in either order.

$4 + m$ is the same as $m + 4$

♦ **Complete each equation. Show the values in the other order if you can.**

a. $5 + j =$

e. $m - 4 =$

b. $4 + 8 =$

f. $16 - a =$

c. $7 - 3 =$

g. $p + 3.5 =$

d. $q + \dfrac{2}{3} =$

Part 2 ⟩ Oral Practice

a. $j - 36 =$

b. $j + 300 =$

c. $j + \dfrac{1}{2} =$

① $j - 12 =$ ▮
$ + 12$
—————
$j \quad =$ ▮

②
$7 + p =$ ▮
$- 7$
—————
$p =$ ▮

◆ **Change each side to get rid of the number that is added or subtracted. Write what's left on the side. Remember the = sign.**

a. $j - 36 =$ ▮

———————

▮

b. $16 + r =$ ▮

———————

▮

c. $j + 300 =$ ▮

———————

▮

d. $t - \dfrac{1}{2} =$ ▮

———————

▮

e. $12 + y =$ ▮

———————

▮

Lesson 20

Part 1 › Changing Both Sides of an Equation

> ◆ If you change **one** side, you must change the other side in the same way.

◆ **For each item, change both sides of the equation in the same way. Figure out what the letter equals.**

a. $k - 2 = 23$

d. $t - 30 = 1$

b. $17 + r = 24$

e. $\dfrac{3}{5} + y = \dfrac{5}{5}$

c. $g + 13 = 17$

f. $m - 100 = 256$

Lesson 21

Part 1 For each item, change both sides of the equation in the same way. Figure out what each letter equals.

a. $5 + b = 22$

c. $q - 1.5 = 3$

e. $n - 10 = 37$

b. $19 + k = 30$

d. $f - \dfrac{1}{3} = \dfrac{1}{3}$

f. $p + \dfrac{5}{9} = \dfrac{12}{9}$

Lesson 22

Part 1 Complete each equation.

a. $\boxed{} \left(\dfrac{2}{95} \right) = \boxed{}$

c. $18 \left(\dfrac{}{} \right) = 1$

e. $208 \left(\dfrac{}{} \right) = d$

b. $r \left(\dfrac{}{} \right) = z$

d. $\boxed{} \left(\dfrac{v}{x} \right) = \boxed{}$

Lesson 24

◆ Any value turned upside down is a value's reciprocal.

◆ The reciprocal of $\frac{2}{3}$ is $\frac{3}{2}$.

◆ Any value multiplied by its reciprocal = 1.

$$\frac{5}{3} \times \frac{3}{5} = \frac{15}{15} = 1$$

$$\frac{5}{3} \times \frac{3}{5} = \frac{3}{3} \times \frac{5}{5}$$
$$= 1 \times 1 = 1$$

Part 2 > Multiplying by the Reciprocal

$$4 \times j = 20$$

$$\left(\frac{1}{4}\right) 4 \times j = 20 \left(\frac{1}{4}\right) \quad \blacktriangleleft \text{ Multiply both sides by the reciprocal.}$$

$$j = \frac{20}{4} \quad \blacktriangleleft \text{ Do the multiplication.}$$

$$\boxed{j = 5} \quad \blacktriangleleft \text{ Write the simple equation.}$$

◆ **Work each item.**

a. $\quad \frac{3}{5} \times m = 6$

c. $\quad k \times 6 = 18$

b. $\quad 11 = g \times \frac{7}{2}$

d. $\quad \frac{1}{8} \times d = 2$

For each problem, simplify, then multiply to figure out the answer.

a. $16\left(\dfrac{3}{6}\right) =$ 　　　

c. $10\left(\dfrac{14}{5}\right) =$ 　　　

e. $7\left(\dfrac{12}{4}\right) =$ 　　　

b. $15\left(\dfrac{18}{10}\right) =$ 　　　

d. $11\left(\dfrac{8}{22}\right) =$ 　　　

Lesson 25

Part 1 **For each problem, simplify, then multiply to figure out the answer.**

a. $\dfrac{5}{8} \times \dfrac{2}{15} =$ 　　　

b. $10\left(\dfrac{14}{4}\right) =$ 　　　

c. $\dfrac{3}{5}\left(\dfrac{20}{6}\right) =$

Lesson 26

$$\frac{1}{9} \times j = \frac{j}{9}$$

$$\frac{6}{7} \times b = \frac{6b}{7}$$

◆ Make a fraction from everything the letter is multiplied by.

◆ Then multiply by the reciprocal.

◆ **Solve each equation.**

a. $\dfrac{5r}{10} = 2$

c. $\dfrac{5}{8} = \dfrac{k}{2}$

b. $\dfrac{h}{11} = 8$

d. $6 = \dfrac{9n}{2}$

Part 2 ⟩ Fraction Simplification

$$\frac{12b}{3b} = \frac{12 \times \cancel{b}^1}{3 \times \cancel{b}} = \boxed{4}$$

◆ If there's a letter over the same letter, you can simplify because that part = 1.

◆ **For each problem, simplify, then multiply to figure out the answer.**

a. $6g\left(\dfrac{2}{9g}\right) =$

d. $\dfrac{5}{3}\left(\dfrac{4p}{7p}\right) =$

b. $15\left(\dfrac{2b}{5c}\right) =$

e. $3b\left(\dfrac{5}{15b}\right) =$

c. $24r\left(\dfrac{7m}{6r}\right) =$

Lesson 28

Part 1 > Simplifying Problems

◆ You've simplified problems that have 2 values multiplied.

$$\frac{4}{\cancel{5}} \times \frac{\cancel{10}^{\,\cancel{5}^{\,1}}}{\cancel{6}_3} = \frac{4}{3}$$

◆ Don't confuse those with problems that add or subtract.

$$\frac{4}{5} + \frac{\cancel{10}^{\,5}}{\cancel{6}_3} = \blacksquare$$

◆ If you **multiply** fractions, it's possible to simplify **across** fractions.

◆ If you **add or subtract,** you **cannot** simplify across fractions.

◆ **Show any simplification that is possible. Do not work the problems.**

a. $\dfrac{12}{3} + \dfrac{5}{15} = \blacksquare$

c. $\dfrac{12}{3}\left(\dfrac{5}{15}\right) = \blacksquare$

e. $9\left(\dfrac{10}{6}\right) = \blacksquare$

b. $\dfrac{7}{3} + \dfrac{9}{14} = \blacksquare$

d. $5 + \dfrac{10}{15} = \blacksquare$

Lesson 30

Part 1 > **For each item, circle the number that is farther from zero.**

a. −3 +1

d. +88 −87

b. +16 −21

e. +1.5 −9.0

c. $-\dfrac{2}{3}$ $+\dfrac{5}{3}$

f. +4 $-\dfrac{4}{3}$

Step ①

$3j + 9 = 21$
$\underline{-9\quad -9}$
$3j = 12$ $\longrightarrow$

Step ②

$\left(\dfrac{1}{3}\right)3j = 12\left(\dfrac{1}{3}\right)$

$\boxed{j = 4}$

a. $16 + 4r = 28$

c. $\dfrac{7}{5} = \dfrac{1}{2}m + \dfrac{4}{5}$

e. $18 = \dfrac{4}{3}r + 10$

b. $\dfrac{2}{3}p - 90 = 6$

d. $4v - 7 = 5$

Part 3 Make a point on the coordinate system for each description.
Label each point.

Point A $x = 5$, $y = 7$
Point B $x = 8$, $y = 10$
Point C $x = 0$, $y = 2$
Point D $x = 3$, $y = 5$

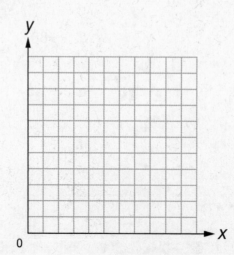

Lesson 31

Part 1 Make a point on the coordinate system for each description. Label each point.

Point A $x = 10,$ $y = 6$

Point B $x = 5,$ $y = 1$

Point C $x = 8,$ $y = 4$

Point D $x = 9,$ $y = 5$

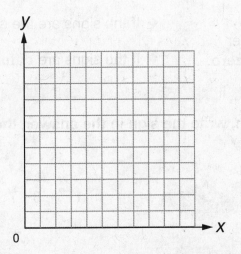

Part 2 First, figure out the sign in the answer. Then figure out the number part.

a. $-6 - 9$ =

d. $+ 15 + 3$ =

b. $- 13 + 11$ =

e. $+ 15 - 20$ =

c. $- 26 + 30$ =

f. $+ 15 - 3$ =

Lesson 32

- ◆ The sign in the **answer** is the sign of the number that is **farther from zero**.

- ◆ If the signs are the **same**, you **add**.

- ◆ If the signs are **different**, you **subtract**.

◆ **For each problem, write the sign in the answer, then figure out the number part.**

a. $+15 - 19 =$

f. $+8 - 7 =$

b. $-11.5 - 6.3 =$

g. $-11.25 + 6.05 =$

c. $+\dfrac{3}{5} + \dfrac{1}{5} =$

h. $+\dfrac{6}{2} - \dfrac{1}{2} =$

d. $-110 + 150 =$

i. $+44 + 14 =$

e. $-\dfrac{7}{11} - \dfrac{2}{11} =$

j. $-1.3 - 3.0 =$

Complete the table. Then draw a line through the points shown.

	x	y
A	3	
B	5	
C	1	
D	6	

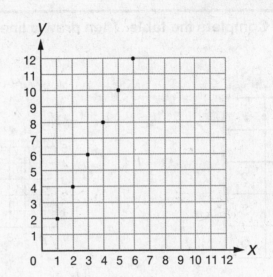

Make a point for each description. Write the letter for each point.

A $x = 4$, $y = 4$

B $x = 8$, $y = 0$

C $x = 10$, $y = 3$

D $x = 0$, $y = 6$

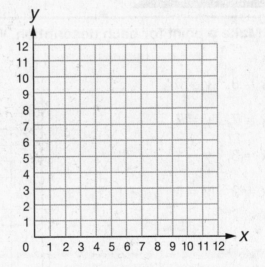

Lesson 33

Part 1 Complete the table. Then draw a line through the points shown.

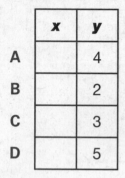

	x	y
A		4
B		2
C		3
D		5

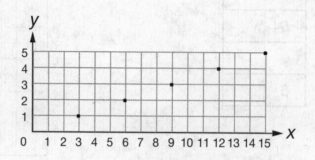

Independent Work

Part 2 Make a point for each description. Write the letter for each point.

A $x = 8$, $y = 0$

B $x = 7$, $y = 7$

C $x = 3$, $y = 1$

D $x = 9$, $y = 5$

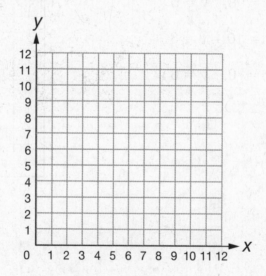

Lesson 34

Part 1 Check each point. If a point is shown incorrectly, plot it again.
Then draw the correct line.

	x	y
M	2	1
J	4	2
D	8	4
F	12	6

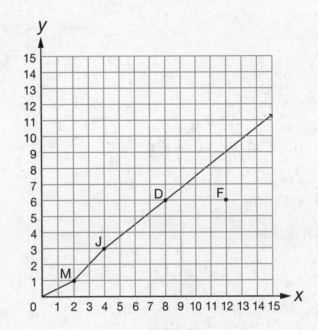

Lesson 35

Part 1 Plot points A and B. Draw the line. Then plot points C through E
and complete the table.

	x	y
A	2	6
B	4	12
C		9
D	1	
E	5	

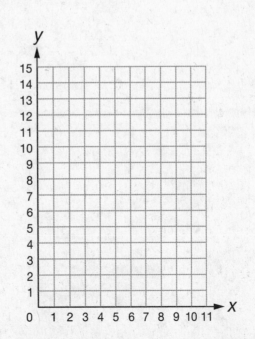

First multiply to remove the parentheses.
Then solve for the letter.

a. $4z - 3(7) = 3$

d. $8q - 2 = 12\left(\dfrac{1}{6}\right)$

b. $8(4) = 20 + 3p$

e. $9 + 16(1) = 10v$

c. $\dfrac{3}{2}(5) - \dfrac{3}{2} = 6j$

Part 3 Make a point for each description. Write the letter for each point.

A $x = 6, \quad y = 0$

B $x = 2, \quad y = 10$

C $x = 1, \quad y = 1$

D $x = 5, \quad y = 3$

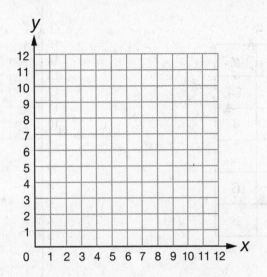

Lesson 36

Part 1 First multiply to remove the parentheses.
Then solve for the letter.

a. $6t + 11(2) = 31$ c. $7(2) = 10 + \frac{1}{2}r$ e. $\frac{2}{7}w = 5(2) - 6$

b. $4m = 5\left(\frac{1}{2}\right) - \frac{3}{2}$ d. $5(4) - 2 = 3f$ f. $5q - 9(4) = 14$

Plot points A and B. Draw the line.
Then plot points C through E and complete the table.

	x	y
A	7	14
B	2	4
C	5	
D		6
E		8

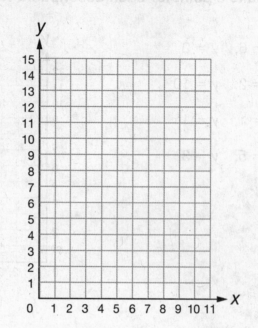

Complete the table.

Fraction	Decimal	%
	.06	
		403%
$\frac{113}{100}$		
		5%
	4.31	

	x	y
A	8	0
B	5	6
C	6	5

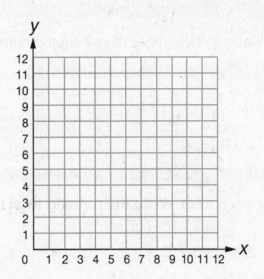

Lesson 37

Part 1 ▶ Complete the table. Then plot 3 points and draw the line.
Then plot the remaining points.

Function			
x	+ 3	=	y
A	6		
B	9		
C	0		
D	2		
E	7		
F	4		

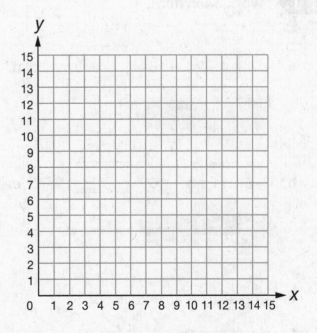

◆When you combine more than 2 signed numbers, you can keep a running total.

$+ 6 + 11 - 13 - 10 =$ $\boxed{-6}$

[+ 17] [+ 4]

$- 6 - 4 - 3 + 10 =$ $\boxed{-3}$

[– 10] [– 13]

◆ **Work each item. Write the running total below each problem.**

a. $-2 + 8 - 5 + 6 =$ ☐

b. $+ 6 - 7 - 4 + 8 =$ ☐

Lesson 38

Part 1 > Work each item.

a. $- 7 + 4 - 20 + 18 =$ ☐

d. $+ 15 - 8 - 10 + 7 - 1 =$ ☐

b. $+ 5 - 11 - 3 + 16 =$ ☐

e. $- 100 + 25 - 75 + 50 + 50 =$ ☐

c. $- 60 + 10 + 40 - 20 + 50 =$ ☐

Function			
x	$\left(\dfrac{1}{2}\right)$	$=$	y
A	6		
B	2		
C	10		
D	8		
E	12		

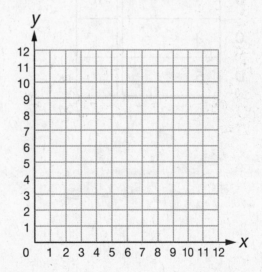

Independent Work

Part 3 Make a point for each description. Write the letter for each point.

A $x = 9$, $y = 1$

B $x = 3$, $y = 5$

C $x = 0$, $y = 9$

Lesson 39

Part 1 > Function Tables

Function		
x ■ = y		
5	(2)	10
5	+ 5	10

Function		
x ■ = y		
3	■	12
3	■	12

Function		
x ■ = y		
6	■	18
6	■	18

◆ **Figure out the correct function. Complete each table.**

①

Function		
x (2) = y		
x + 4 = y		
A	4	8
B	6	10
C	3	
D	5	

②

Function		
x (2) = y		
x + 3 = y		
A	3	6
B	8	16
C	5	
D	7	

Part 2 Complete the table.

Fraction	Decimal	%
$\dfrac{3}{100}$		
		800%
	7.42	
		3%
$\dfrac{500}{100}$		

Lesson 40

Part 1 ▶ Complete each table to show 2 possible functions for the pair of values shown.

a.
Function	
x	= y
x	= y
11 ▮	22

d.
Function	
x	= y
x	= y
6 ▮	30

b.
Function	
x	= y
x	= y
9 ▮	27

e.
Function	
x	= y
x	= y
3 ▮	12

c.
Function	
x	= y
x	= y
5 ▮	5

Figure out the correct function. Complete the table.

Function		
x	**= y**	
x	**= y**	
7		14
0		7
2		
13		
9		

Part 3 Label the axes. Write the description for each point. The description for A is given.

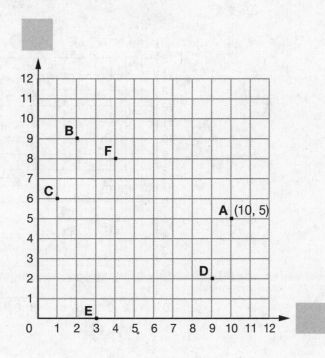

Part 4 — Label the axes. Plot and label each point on the coordinate system.

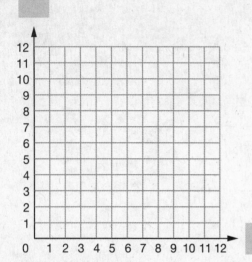

A	(0, 7)	**D**	(5, 0)
B	(2, 9)	**E**	(12, 2)
C	(10, 1)	**F**	(3, 6)

Part 5 — Label the axes. Plot and label each point on the coordinate system.

A (3, 9)

B (5, 15)

C (1, 3)

D (4, 12)

E (2, 6)

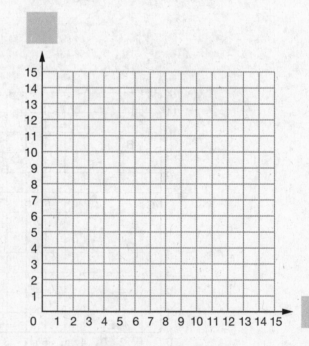

Lesson 41

Part 1 → **Write 2 functions for each pair of values.**

Sample Problem:

Function	
$x \left(\frac{3}{4}\right) = y$	
$x \quad -1 = y$	
4 ▮	3

a.

Function	
$x \quad = y$	
$x \quad = y$	
5 ▮	9

b.

Function	
$x \quad = y$	
$x \quad = y$	
18 ▮	11

c.

Function	
$x \quad = y$	
$x \quad = y$	
7 ▮	4

d.

Function	
$x \quad = y$	
$x \quad = y$	
9 ▮	20

e.

Function	
$x \quad = y$	
$x \quad = y$	
8 ▮	1

Part 2	Label the axes. Write the description for points A–E. Plot points F–J on the coordinate system.

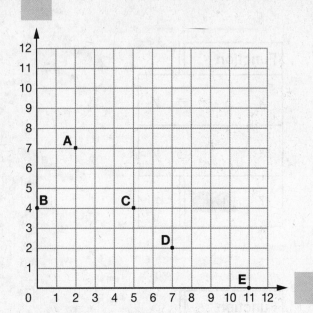

F   (7, 11)

G   (2, 5)

H   (0, 9)

I   (12, 1)

J   (3, 0)

> Independent Work <

Part 3	Cross out the function that is wrong. Complete the table.

Function	
x + 4 = y	
x (3) = y	
2	6
4	8
5	
9	

Lesson 42

Part 1 ⟩ Combining Like Terms

$$-4m + 6m + 11m \qquad -7k + 6b - 14 + 2p - 4b$$

$$+13m \qquad\qquad -7k + 2b - 14 + 2p$$

◆ **Rewrite each expression with like terms combined.**

a. $4p - 2m + 3 - 6p - 7$

c. $-k + 5 + 2k + 5k - 11$

b. $4 - 3q + 7v - 6q - 3$

d. $-12j + 10 + 8j - j - 9h$

a.

Function		
x	=	**y**
x	=	**y**
3	▉	16

d.

Function		
x	=	**y**
x	=	**y**
2	▉	3

b.

Function		
x	=	**y**
x	=	**y**
9	▉	7

e.

Function		
x	=	**y**
x	=	**y**
6	▉	1

c.

Function		
x	=	**y**
x	=	**y**
15	▉	4

Independent Work

Part 3

Cross out the function that is wrong. Complete the table.

Function	
x (3) = **y**	
x + 8 = **y**	
4	12
10	30
7	
2	
11	

Lesson 43

Part 1 Write the coordinates next to each point. You do not need a plus sign for positive *x* or *y* values.

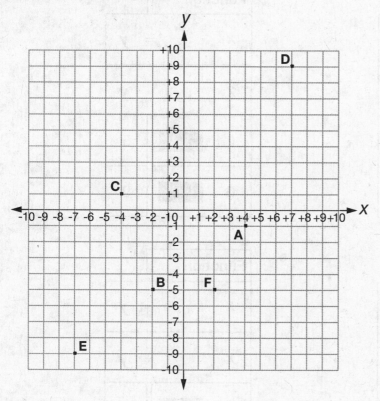

Part 2 Rewrite each expression with like terms combined.

a. $15 - 3c + 1 + 8d + 7c - 10$

d. $8 - 4n + 10 + n - 9$

b. $20w - 34 - 10w + 8z - 5z$

e. $+ 4p - 2p + 28 - p - 11 - m$

c. $-15r + r - 7f - 4r + 11$

Complete 2 functions for each table.
Cross out the function that does not work for both rows.

a.

Function		
x	=	y
x	=	y
5	■	4
10	■	8

d.

Function		
x	=	y
x	=	y
2	■	5
6	■	9

b.

Function		
x	=	y
x	=	y
6	■	5
12	■	11

e.

Function		
x	=	y
x	=	y
8	■	6
12	■	9

c.

Function		
x	=	y
x	=	y
1	■	5
7	■	35

Lesson 44

$$-3a + 11 + 7a - 2 = 17 \quad \blacktriangleleft \text{ Combine like terms.}$$
$$4a \qquad + 9 \qquad = 17 \quad \blacktriangleleft \text{ Copy the rest of the equation.}$$

◆ **Below each item, write an equation that shows the simplified expression with the letter terms and number terms combined. Then solve for the letter.**

a. $5r - 2 + 3 + 8r - 6r = 3$

c. $-\dfrac{2}{3}t + \dfrac{7}{3}t - \dfrac{1}{3}t + 15 - 12 = 9$

b. $6 - 2b - 10 + 7b = 1$

d. $14 + 8j + 2 + 12j - 7j = 18$

a.

Function		
x	=	*y*
x	=	*y*
7		4
14		11
3		
6		
10		

b.

Function		
x	=	*y*
x	=	*y*
2		8
5		20
3		
1		
6		

c.

Function		
x	=	*y*
x	=	*y*
8		10
4		5
12		
16		
0		

Write the coordinates next to each point. You do not need a plus sign for positive *x* or *y* values.

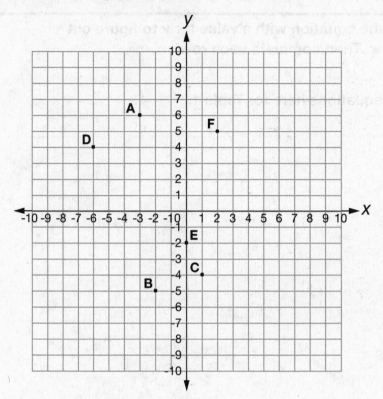

Part 4 Cross out the function that is wrong. Complete the table.

Function	
x + 12 = *y*	
x (3) = *y*	
6	18
7	21
10	
4	

Lesson 45

Part 1 For each table, write the equation with a value for *y* to figure out the *x* value in each row. Then complete each row.

Table 1

Function		
x $\left(\dfrac{y}{x}\right)$ =		y
x $\left(\dfrac{2}{3}\right)$ =		y
A	$\left(\quad\right)$	4
B	$\left(\quad\right)$	10
C	$\left(\quad\right)$	8
D	$\left(\quad\right)$	14

Solve equations here for Table 1.

Table 2

Function		
x $\left(\dfrac{y}{x}\right)$ =		y
	$\left(\dfrac{7}{4}\right)$	
A		7
B		21
C		14
D		35

Solve equations here for Table 2.

Part 2 > **Write the coordinates next to each point.**

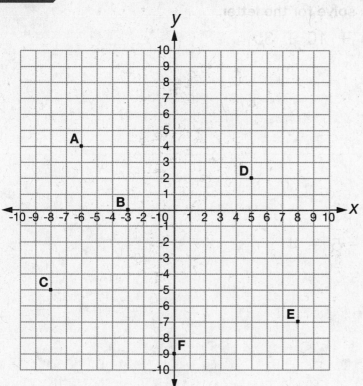

Below each item, write an equation with the letter and number terms combined. Then solve for the letter.

a. $26 = -3 + 13 - p + 10 + 3p$

b. $1 = +3t - t + 8t - 5 + 4$

c. $16 = -5r - 12 + 3 + 4r + 6r$

d. $0 = -4 + 10b - 2b - 6b - 12$

Part 4 ▶ Make a point for each pair of coordinates. Write the letter for each point.

E   (5, 7)

F   (4, 8)

G   (1, 10)

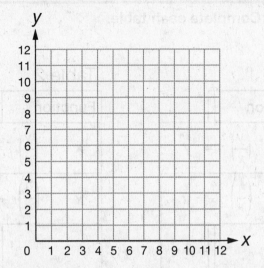

Part 5 ▶ Figure out the correct function. Cross out the function that is wrong. Complete the table.

Function		
x	=	y
x	=	y
6		9
2		3
16		
12		
20		

Lesson 46

Part 1 Complete each table.

Table 1

	Function		
	x	$\left(\frac{y}{x}\right)$ =	y
	x	$(-)$ =	y
A	2		3
B	26		
C			21

Table 2

	Function		
	x	$\left(\frac{y}{x}\right)$ =	y
	x	$(-)$ =	y
A	1		5
B			15
C	18		

Table 3

	Function		
	x	$\left(\frac{y}{x}\right)$ =	y
	x	$(-)$ =	y
A	2		1
B			2
C	10		

Independent Work

Part 2 Round each value to a whole number, 10ths, 100ths.

		Whole Number	Tenths	Hundredths
Sample	12.0549	12	12.1	12.05
	.8624			
	4.0018			
	2.356			
	10.953			

50 *Lesson 46*

Make a point for each pair of coordinates. Write the letter for each point.

E $x = 10$, $y = 9$

F $x = 6$, $y = 5$

G $x = 0$, $y = 5$

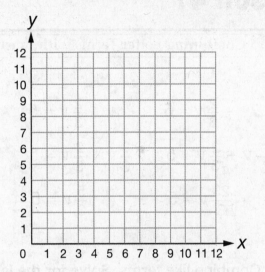

Part 4 Figure out the correct function. Complete the table.

Function	
x $=$ y	
x $=$ y	
1	4
8	11
5	
0	
3	

Lesson 47

Combining Letter Terms with Fractions

$$7v + \frac{2}{3}v = 5 \qquad v + \frac{2}{3}v = 5$$

$$\frac{21}{3}v + \frac{2}{3}v = 5 \qquad \frac{3}{3}v + \frac{2}{3}v = 5$$

$$\frac{23}{3}v = 5 \qquad \frac{5}{3}v = 5$$

◆ **Combine like terms. Solve for the letter.**

a. $\frac{9}{5}v + v = 28$

c. $0 = r + 2 - 6 - \frac{1}{5}r$

b. $3j - \frac{1}{2}j + 11 - 10 = 21$

Table 1

Function		
x	$\left(\dfrac{y}{x}\right) = y$	
x	$(-) = y$	
A	4	3
B		21
C	20	

Table 2

Function		
x	$\left(\dfrac{y}{x}\right) = y$	
x	$(-) = y$	
A	2	6
B	12	
C		9
D	1	

Table 3

Function		
x	$\left(\dfrac{y}{x}\right) = y$	
x	$(-) = y$	
A	9	3
B		5
C	6	

Part 3 Make a point for each pair of coordinates. Write the letter for each point.

E (10, 9)

F (6, 5)

G (0, 10)

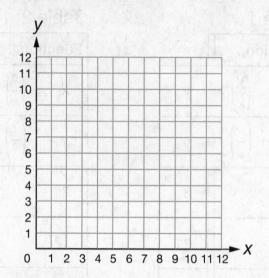

Part 4

Figure out the correct function. Complete the table.

Function		
x	=	y
x	=	y
A	2	4
B	6	12
C	0	
D	1	
E	4	

Part 5

Round each value to a whole number, 10ths, and 100ths.

	Whole Number	Tenths	Hundredths
40.386	40	40.4	40.39
.6945			
3.064			
12.709			

Lesson 48

Part 1 **Work each item. First figure out the sign. Then multiply.**

a. $- 5 \; (-2) =$

b. $+ 10 \; (-8) =$

c. $- 2 \; (+3) =$

d. $+ \dfrac{1}{2} \left(+ \dfrac{3}{5} \right) =$

e. $- 3 \; (-8) =$

f. $- 1.5 \; (+4) =$

Part 2 **Solve for each letter.**

a. $5g = 16 - 3g$

b. $3p + 7 = 10p$

c. $10 + 5k - 4 = 7k$

d. $4v = 14 - 2v - 2$

e. $1 - 3m + 5 = 6m$

Part 3 Complete each table.

Table 1

Function		
x $\left(\dfrac{y}{x}\right)$ = y		
x $\left(-\right)$ = y		
A	5	2
B	7	
C		5
D		8

Table 2

Function		
x $\left(\dfrac{y}{x}\right)$ = y		
x $\left(-\right)$ = y		
A	4	1
B		$\dfrac{1}{2}$
C	24	
D	3	

Part 4

Round each value to the nearest hundredth.

a. 13.027 _____

b. 0.1335 _____

c. 4.3250 _____

d. .146 _____

e. 100.024 _____

Part 5

Figure out the correct function. Complete the table.

Function	
x = y	
x = y	
2	6
4	8
10	
1	
8	

Lesson 49

Part 1 Work each item.

a. Figure out the correct function. Cross out the function that is wrong, then complete the table.

Function		
$x\ (\quad)\ =\ y$		
$x\qquad=\ y$		
A	10	5
B	5	0
C	13	
D	6	
E	9	

b. Complete the function. Then complete the table.

Function		
$x\ \left(\frac{y}{x}\right)\ =\ y$		
$x\ \left(-\right)\ =\ y$		
A	6	1
B	18	
C		6
D	72	
E		18

Part 2 Round each value to a whole number, 10ths, and 100ths.

	Whole Number	Tenths	Hundredths
4.308			
10.365			
0.926			

Lesson 50

Part 1 Complete the table.

Function		
x $\left(\dfrac{y}{x}\right) = y$		
x $(-) = y$		
A	4	5
B		25
C	12	
D		10
E	28	

Part 2 Round each value to a whole number, 10ths, and 100ths.

	Whole Number	Tenths	Hundredths
3.546			
2.099			
6.340			

Lesson 51

Independent Work

Part 1 Round each value to a whole number, 10ths, and 100ths.

	Whole Number	Tenths	Hundredths
47.830			
.0237			
5.0308			
16.478			

Lesson 52

Part 1 Complete the table. Write the y value as a whole number or as a simplified fraction.

	Function
	$y = \left(\dfrac{y}{x}\right) x$
	$y = \left(\dfrac{3}{2}\right) x$
a.	4
b.	1
c.	10
d.	0
e.	6

Below each item, rewrite the equation with like terms combined. Then solve for the letter.

a. $7k = 30 + k - 2k + 2$

d. $\frac{2}{5}r = 13 - 4r - 2$

b. $11 - w + 4 + 2w = 4w$

e. $4p - 13 - 3p + 18 = 19$

c. $12n - 10n + 5n = -\frac{3}{2} + \frac{10}{2}$

Lesson 53

Part 1

Complete the table. Show the y values as simplified values.

Function		
$y = \left(\frac{y}{x}\right)$	x	
$y = \left(\frac{1}{3}\right)$	x	
a.	3	
b.	15	
c.	30	
d.	1	
e.	0	

Part 2

Round each value to a whole number, 10ths, and 100ths.

	Whole Number	Tenths	Hundredths
.0538			
2.609			
3.09			
5.828			

Part 3

Figure out the correct function. Complete the table.

Function		
x	$\left(\frac{y}{x}\right) = y$	
x	$\left(\frac{\ }{\ }\right) = y$	
7	8	
5	3	
2		
	56	

Part 1 ▶ Complete the table. Show the y values as simplified values.

Function	x
$y = \left(\dfrac{y}{x}\right)x$ $y = \left(\dfrac{3}{4}\right)x$	
A	1
B	8
C	40
D	100

Independent Work ▶

Part 2 ▶ Figure out the correct function. Complete each table.

Table 1

Function	
x = y	
x	y
6	5
11	10
7	
5	
9	
20	

Table 2

Function	
$x\left(\dfrac{y}{x}\right)$ = y	
x	y
6	2
5	18
10	
	24

Lesson 55

point

A $y = \left(\dfrac{9}{5}\right)x$ ( )

B $y = \left(\dfrac{3}{7}\right)x$ ( )

C $y = \left(\dfrac{1}{6}\right)x$ ( )

D $y = \left(\dfrac{5}{4}\right)x$ ( )

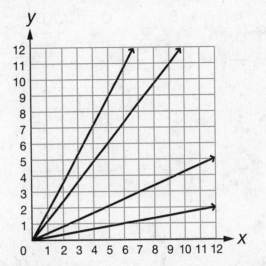

Independent Work

Part 2 Figure out the correct function. Complete the table.

Function		
x $\left(\dfrac{y}{x}\right)$	=	y
x $\left(\dfrac{\ }{\ }\right)$	=	y
4		3
		12
9		
		9

Lesson 56

Use the coordinates for each point to complete the equivalent fraction equation.

Line 1 $\dfrac{y}{x}$ =

A	B	C	D	E
——	——	——	——	——

Line 2 $\dfrac{y}{x}$ =

A	B	C	D	E
——	——	——	——	——

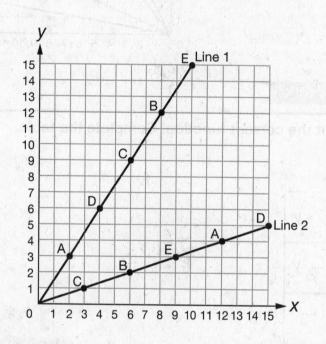

Independent Work

Part 2 Round each value to a whole number, 10ths, and 100ths.

	Whole Number	Tenths	Hundredths
18.295			
3.006			
27.083			
.658			

Lesson 57

Part 1 Complete the table.
Use the equation $y = \frac{6}{5}x$ to figure
out the missing x values.

		$y = \frac{6}{5}x$
A	18	
B	6	
C	12	
D	24	

Part 2 Plot the points A through D on the coordinate system.
Show the coordinates next to each point.
Draw a line through those points.

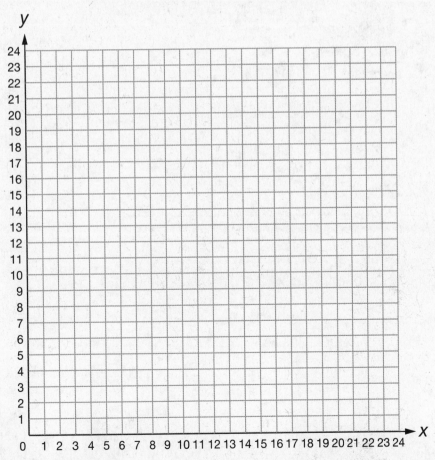

Part 3 Round each value to a whole number, 10ths, and 100ths.

	Whole Number	Tenths	Hundredths
3.008			
.3051			
24.099			
6.304			

Lesson 58

Part 1   Complete the table. Plot points A through D. Draw the line.

$y = \frac{3}{5} x$		
A	6	
B		20
C		15
D	15	

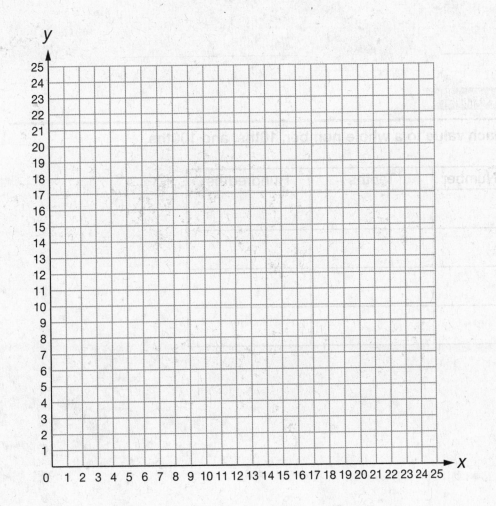

Part 2 — Order of Operations

$$-2\,(+4) - 1 + 8\,(-2) = \blacksquare$$

$$-8 \quad -1 \quad -16 \;=\; \blacksquare$$

♦ First remove parentheses.

♦ Then combine.

a. $+6 - 2\,(-3) + 4\,(-1) - 5 = \blacksquare$

_____ =

c. $-7\,(-1) + 12 - 2\,(-9) = \blacksquare$

_____ =

b. $+2\,(-5) - 8 - 3\,(3) + 6 = \blacksquare$

_____ =

d. $-4\,(+6) - 8\,(7) - 10 = \blacksquare$

_____ =

Independent Work

Part 3 Round each value to a whole number, 10ths, and 100ths.

	Whole Number	Tenths	Hundredths
16.099			
34.638			
17.906			
5.751			

Lesson 59

Part 1

For each line, write the fraction for the slope. Then make 2 more points on the line.

Part 2

Rewrite each fraction. First show a base with a positive exponent. Then show a base with a negative exponent.

Sample
$$\frac{4 \times 4 \times 4}{4 \times 4 \times 4 \times 4 \times 4} = \frac{1}{4^2} = 4^{-2}$$

a. $\dfrac{7 \times 7 \times 7 \times 7 \times 7}{7 \times 7} =$

b. $\dfrac{8 \times 8 \times 8}{8 \times 8 \times 8 \times 8 \times 8} =$

c. $\dfrac{10}{10 \times 10 \times 10 \times 10 \times 10} =$

d. $\dfrac{m \times m \times m \times m \times m \times m \times m}{m \times m} =$

Work each problem. Multiply to remove the parentheses. Then combine values to figure out the answer.

a. $+8 - 3(4) + 4(-6) - 10$ $=$ ▮

_____ $=$

b. $+11 + 2(-6) - 4(-2)$ $=$ ▮

_____ $=$

c. $-4(-3) - 4 - 3 + 17$ $=$ ▮

_____ $=$

d. $-18 - 1(+6) + 6 - 18$ $=$ ▮

_____ $=$

e. $-3(+2) - 6 + 12(-1) + 20$ $=$ ▮

_____ $=$

Part 4 > Round each value to a whole number, 10ths, and 100ths.

	Whole Number	Tenths	Hundredths
17.006			
2.347			
.872			
22.265			

Part 5 > Complete the table. Plot and label each point. Show the coordinates. Then draw the line.

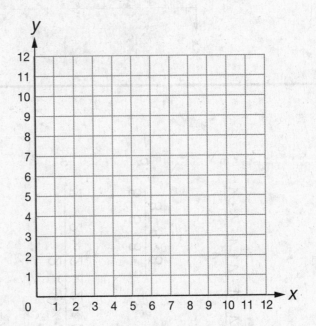

	Function $y = \left(\dfrac{2}{3}\right) x$
A	12
B	6
C	6
D	2

Lesson 60

Write the slope for each line.

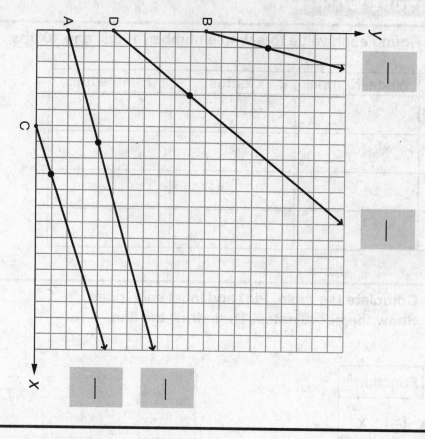

Rewrite each fraction. First show a base number with a positive exponent. Then show a base number with a negative exponent.

a. $\dfrac{5 \times 5 \times 5 \times 5}{5} = $ ⬚ $ = $ ⬚

b. $\dfrac{r \cdot r}{r \cdot r \cdot r} = $ ⬚ $ = $ ⬚

c. $\dfrac{2 \times 2 \times 2}{2 \times 2 \times 2 \times 2 \times 2} = $ ⬚ $ = $ ⬚

d. $\dfrac{8 \cdot 8 \cdot 8 \cdot 8}{8 \cdot 8 \cdot 8} = $ ⬚ $ = $ ⬚

e. $\dfrac{v}{v \cdot v \cdot v \cdot v \cdot v} = $ ⬚ $ = $ ⬚

Rewrite each problem below. First remove the parentheses, then combine values to figure out the answer.

a. $-1\,(7) + 9 - 2\,(-2) + 5 = $ ■

d. $+8 - 5\,(-2) + 10\left(+\dfrac{1}{2}\right) + 20\left(-\dfrac{1}{4}\right) = $ ■

b. $+3 + 5\,(-8) - 4\,(+6) = $ ■

e. $-3\,(+2) - 6 + 12\,(-1) + 20 = $ ■

c. $-\dfrac{3}{2}\left(-\dfrac{1}{2}\right) + \dfrac{1}{4} + 1\left(\dfrac{2}{4}\right) - \dfrac{1}{2}\left(+\dfrac{1}{2}\right) = $ ■

Independent Work

Part 4 Round each value to a whole number, 10ths, and 100ths.

	Whole Number	Tenths	Hundredths
.565			
13.086			
5.751			
1.356			

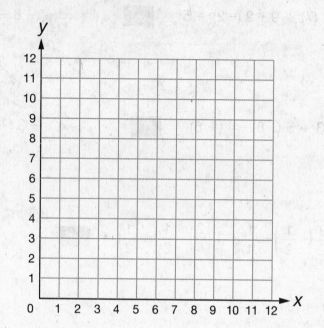

	Function
	$y = \left(\dfrac{1}{2}\right) x$
A	3	
B		12
C	4	
D		4

Lesson 61

Independent Work

Part 1 Complete the table and plot the line. Label each point.

	Function $y = \frac{2}{5} x$	
A	6	
B	4	
C		10
D		5

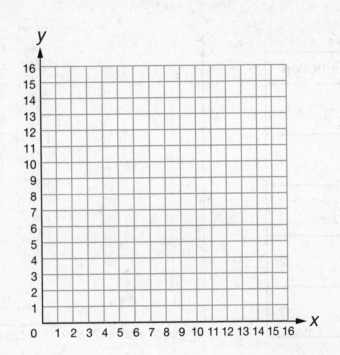

Lesson 63

Part 1 Complete the table and plot the line.

	Function $y = \dfrac{7}{2}\ x$
A	4
B	14	
C	2
D	0

Lesson 64

Independent Work

Part 1 Round each value to a whole number, 10ths, and 100ths.

	Whole Number	Tenths	Hundredths
.7538			
70.046			
3.606			
10.605			

Lesson 69

Part 1 | Simplifying Expressions with Exponents

$m^8\, 5^2\, 5 = 5^3\, m^8 = \boxed{125\, m^8}$

◀ Write the number value first, then the letter value.

$k^{-1}\, 2^3\, b^{-2}\, k^2\, 2^{-7} = 2^{-4}\, kb^{-2} = \boxed{\dfrac{1}{16}\, kb^{-2}}$

◀ If the number has a negative exponent, write it as a fraction.

Part 2 Simplify. Rewrite the expression with a number base first.
Then write the expression with a number that has no exponent.

a. $j^{-3}\, r^3\, 15^3\, r$

b. $7^3\, k^3\, 7^2\, k^{-2}$

c. $b^{-3}\, b\, 4^2\, m^{-1}\, 4^{-5}$

d. $g^{-1}\, n^2\, 10^3\, n\, 10^{-1}$

e. $3^2\, p^3\, p^{-1}\, 3^{-5}\, d^{-10}$

Lesson 70

Draw a point for each line. Then write the equation for each line.

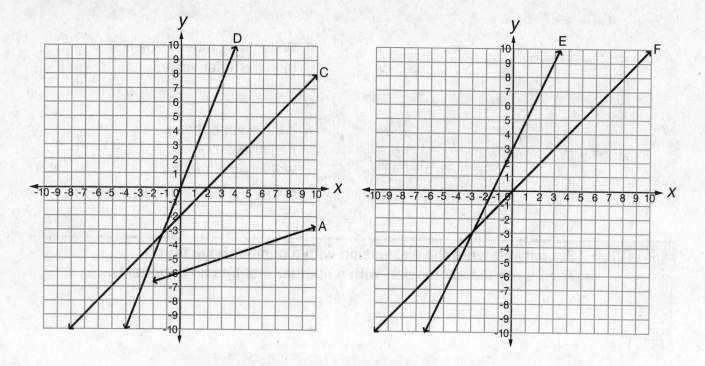

Lesson 71

Part 1 Draw a point for each line. Then write the complete equation for each line.

A

B

C

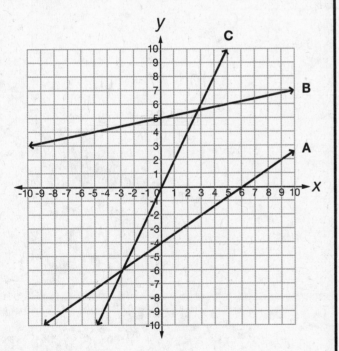

Lesson 72

Part 1 Draw a point for each line. Then write the complete equation for each line.

A

B

C

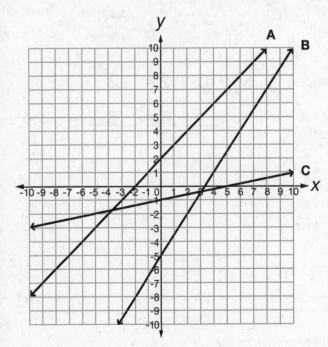

Lesson 73

Part 1 ▸ Write the combined equation below each item. Then solve for the letter.

a. $-3p + 5 = 11$

   $p - 2 = -4$
   _____

c. $-13 + 2r = -5$

   $10 - 3r = -2$
   _____

b. $-4v + 2 = -2$

   $-v + 6 = 5$
   _____

d. $10 - f = 11$

   $-6 + 3f = -9$
   _____

Lesson 74

Part 1 Write the combined equation below each item. Then solve for the letter.

a. $-6 - 2y = -1$

$\underline{5 + 4y = -5}$

c. $-8d + 9 = 5$

$\underline{4d - 3 = -1}$

b. $15 - 5x = 30$

$\underline{10 + 2x = 4}$

d. $2m - 5 = 9$

$\underline{-m + 11 = 4}$

Part 2 Plot a point and write the complete equation for each line.

A

B

C

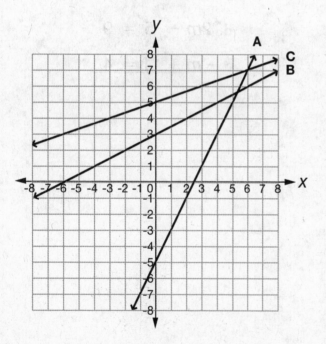

Part 1 Plot a point and write the complete equation for each line.

A

B

C

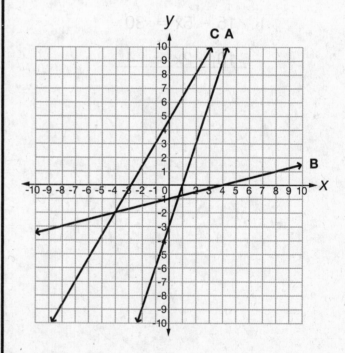

Lesson 76

Independent Work

Part 1 Plot a point and write the complete equation for each line.

A

B

C

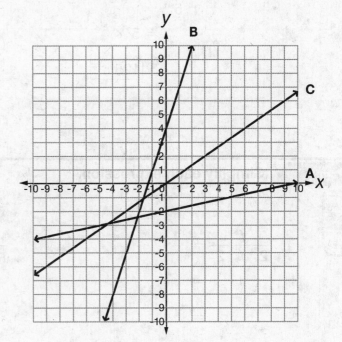

Lesson 77

Part 1 Multiply 1 letter term in each pair so the combined terms disappear.

$$- 4r \left(\boxed{} \right) = +12r$$

$$-12r \quad = \underline{-12r}$$

a. $-8b \quad = -8b$

 $+2b \quad = \underline{}$

b. $5m \quad = 5m$

 $-m \quad = \underline{}$

c. $-3p \quad =$

 $-9p \quad = \underline{}$

d. $+ \ 2g \quad =$

 $+12g \quad = \underline{}$

◆ If you have a fraction with terms multiplied, move all the terms to the **numerator** before you combine exponents.

$$\frac{r^{-2}\,5^2}{r^2\,5^{-1}} = r^{-2}\,5^2\,r^{-2}\,5^1$$

$$= 5^3\,r^{-4}$$

$$= \boxed{125\,r^{-4}}$$

Part 3 ▷ Simplify each expression.

a. $\dfrac{f^3}{2^{-5}\,f^{-2}\,2^3} =$

$=$

$=$

c. $\dfrac{m^4\,3^7}{m^{-2}\,3^9\,m^3} =$

$=$

$=$

b. $\dfrac{v\,10^{-2}}{v^{-2}\,10^2} =$

$=$

$=$

Lesson 78

Part 1 Write the coordinates for each point.

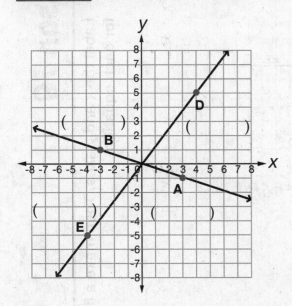

Write 2 equations for each line.

Line at A

Line at B

Line at D

Line at E

Part 2 Simplify each expression.

a. $\dfrac{qm^{-1}\,7^{-2}}{7^{-3}\,q^{-1}\,m^2} =$

$=$

$=$

c. $\dfrac{p^{-3}\,5\,v^{-5}}{v^{-3}\,p^{-1}\,5^{-2}} =$

$=$

$=$

b. $\dfrac{3k^4\,r^{-1}}{k^2\,r^3\,3^5} =$

$=$

$=$

Lesson 79

Label the *x* and *y* axes. Then make a line on the coordinate system for each equation.

A $y = \frac{2}{5}x - 3$ B $y = -\frac{3}{2}x - 1$

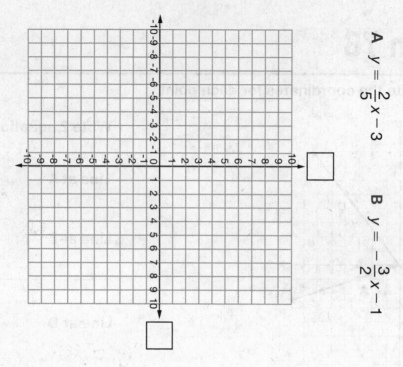

C $y = -\frac{5}{4}x + 8$

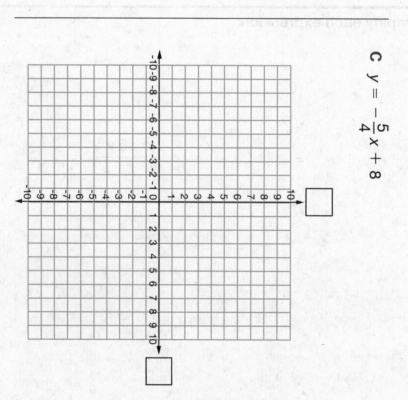

Lesson 80

Part 1 Label the *x* and *y* axes. Then make a line for each equation.

A $y = 3x - 4$

B $y = -x + 5$

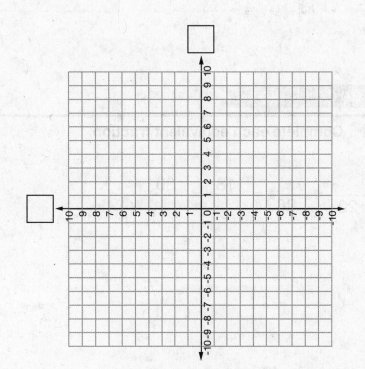

C $y = -\frac{3}{5}x + 2$

D $y = x - 8$

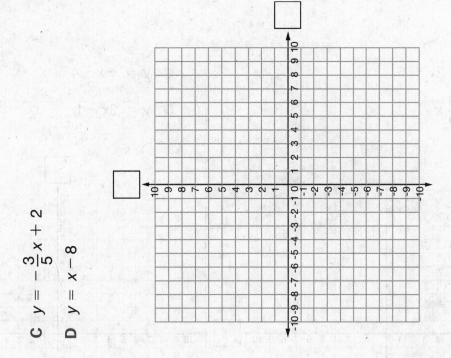

Lesson 81

Label the x and y axes. Then make a line for each equation.

A $y = -2x + 6$

B $y = \dfrac{5}{9}x$

C $y = -x - 4$

D $y = 3x - 1$

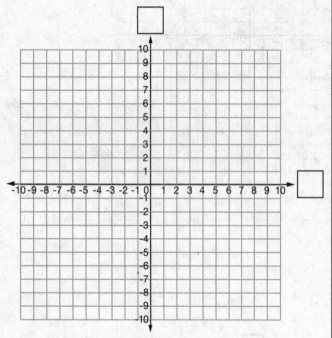

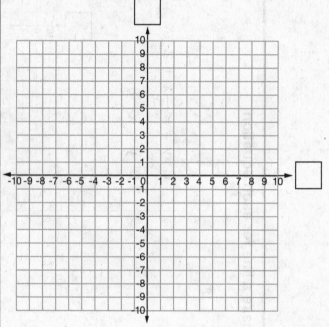

Independent Work

Part 2 **Complete each equivalent fraction.**

a. $\dfrac{3}{4} = \dfrac{}{20} = \dfrac{}{36} = \dfrac{18}{} = \dfrac{30}{}$

Lesson 82

Part 1 Label the *x* and *y* axes. Then make a line for each equation.

A $y = -\dfrac{1}{2}x + 3$

B $y = 4x - 2$

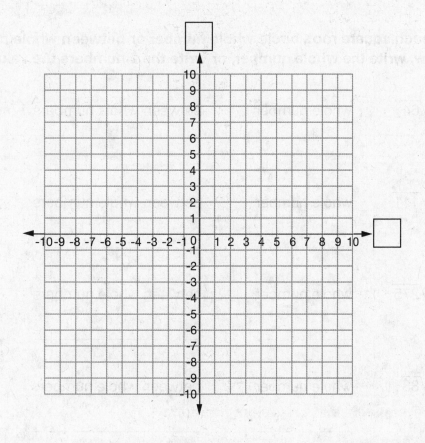

Lesson 83

Part 1 ▷ Whole Number Square Roots, 1–20

$\frac{1}{\sqrt{1}}$	$\frac{2}{\sqrt{4}}$	$\frac{3}{\sqrt{9}}$	$\frac{4}{\sqrt{16}}$	$\frac{5}{\sqrt{25}}$	$\frac{6}{\sqrt{36}}$	$\frac{7}{\sqrt{49}}$	$\frac{8}{\sqrt{64}}$	$\frac{9}{\sqrt{81}}$	$\frac{10}{\sqrt{100}}$

$\frac{11}{\sqrt{121}}$	$\frac{12}{\sqrt{144}}$	$\frac{13}{\sqrt{169}}$	$\frac{14}{\sqrt{196}}$	$\frac{15}{\sqrt{225}}$	$\frac{16}{\sqrt{256}}$	$\frac{17}{\sqrt{289}}$	$\frac{18}{\sqrt{324}}$	$\frac{19}{\sqrt{361}}$	$\frac{20}{\sqrt{400}}$

◆ **For each square root, circle whole number or between whole numbers. Below, write the whole number, or write the 2 numbers the value is between.**

a. $\sqrt{64}$ whole number   between whole numbers

_____   _____

b. $\sqrt{111}$ whole number   between whole numbers

_____   _____

c. $\sqrt{275}$ whole number   between whole numbers

_____   _____

d. $\sqrt{83}$ whole number   between whole numbers

_____   _____

e. $\sqrt{144}$ whole number   between whole numbers

_____   _____

Part 2 Complete each equivalent fraction.

a. $\dfrac{7}{5} = \dfrac{14}{\boxed{}} = \dfrac{\boxed{}}{35} = \dfrac{56}{\boxed{}} = \dfrac{.7}{\boxed{}}$

Part 3 Label the *x* and *y* axes. Then make a line for each equation.

A $y = \dfrac{3}{4}x + 8$

B $y = -2x - 1$

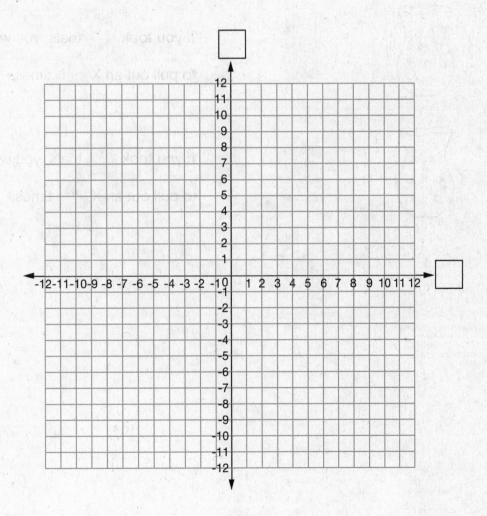

Lesson 84

Part 1 For each item, write the fraction that tells about your chances of pulling an X from the bag. Then complete the sentence that tells about the trials.

a. If you took ___ trials, you would expect to pull out an X ___ times.

b. If you took ___ trials, you would expect to pull out an X ___ times.

c. If you took ___ trials, you would expect to pull out an X ___ times.

d. If you took ___ trials, you would expect to pull out an X ___ times.

1	2	3	4	5	6	7	8	9	10
$\sqrt{1}$	$\sqrt{4}$	$\sqrt{9}$	$\sqrt{16}$	$\sqrt{25}$	$\sqrt{36}$	$\sqrt{49}$	$\sqrt{64}$	$\sqrt{81}$	$\sqrt{100}$

11	12	13	14	15	16	17	18	19	20
$\sqrt{121}$	$\sqrt{144}$	$\sqrt{169}$	$\sqrt{196}$	$\sqrt{225}$	$\sqrt{256}$	$\sqrt{289}$	$\sqrt{324}$	$\sqrt{361}$	$\sqrt{400}$

◆ **For each square root, circle whole number or between whole numbers.
Below, write the whole number or write the 2 numbers the value is between.**

a. $\sqrt{72}$ whole number   between whole numbers

_____   _____

b. $\sqrt{390}$ whole number   between whole numbers

_____   _____

c. $\sqrt{225}$ whole number   between whole numbers

_____   _____

d. $\sqrt{18}$ whole number   between whole numbers

_____   _____

e. $\sqrt{169}$ whole number   between whole numbers

_____   _____

f. $\sqrt{250}$ whole number   between whole numbers

_____   _____

Part 3 Label the *x* and *y* axes. Then make a line for each equation.

A $y = -x + 5$

B $y = -\dfrac{2}{5}x - 5$

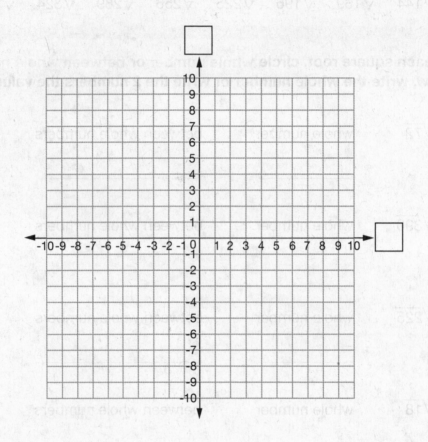

Part 4 Complete each equivalent fraction.

a. $\dfrac{3}{8} = \dfrac{30}{\boxed{}} = \dfrac{12}{\boxed{}} = \dfrac{\boxed{}}{48} = \dfrac{\boxed{}}{64}$

Lesson 85

Part 1 **Write the fraction for each statement where it belongs in the equation.**

a. A person took 15 trials and pulled out a star on 6 of those trials.

objects   trials

$$\frac{\bigstar}{\text{all}} \quad \boxed{} = \boxed{}$$

b. There are 3 stars in the bag and a total of 9 objects in the bag.

objects   trials

$$\frac{\bigstar}{\text{all}} \quad \boxed{} = \boxed{}$$

c. Tom pulled 8 stars from the bag. He took a total of 11 trials.

objects   trials

$$\frac{\bigstar}{\text{all}} \quad \boxed{} = \boxed{}$$

d. There were 8 objects in the bag. 1 of them was a star.

objects   trials

$$\frac{\bigstar}{\text{all}} \quad \boxed{} = \boxed{}$$

e. A person took 12 trials and pulled out a star 4 times.

objects   trials

$$\frac{\bigstar}{\text{all}} \quad \boxed{} = \boxed{}$$

For each equation, plot a point and draw a line that goes through zero.

A $y = \dfrac{1}{5}x$

B $y = \dfrac{6}{7}x$

C $y = -\dfrac{5}{3}x$

D $y = -3x$

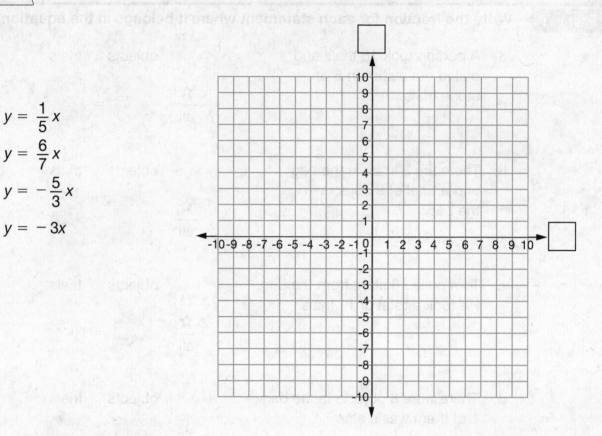

◄ **Independent Work** ►

Part 3 **Complete each equivalent fraction.**

a. $\dfrac{5}{4} = \dfrac{\boxed{}}{24} = \dfrac{60}{\boxed{}} = \dfrac{\boxed{}}{40} = \dfrac{15}{\boxed{}}$

Lesson 86

Part 1 ▸ **Complete each pair of equivalent fractions.**

a.   objects   trials

$\dfrac{☆}{all}$ $\dfrac{3}{5} = \dfrac{\boxed{}}{25}$

d.   objects   trials

$\dfrac{☆}{all}$ $\dfrac{7}{10} = \dfrac{\boxed{}}{100}$

b.   objects   trials

$\dfrac{☆}{all}$ $\dfrac{2}{3} = \dfrac{8}{\boxed{}}$

e.   objects   trials

$\dfrac{☆}{all}$ $\dfrac{1}{2} = \dfrac{15}{\boxed{}}$

c.   objects   trials

$\dfrac{☆}{all}$ $\dfrac{5}{8} = \dfrac{35}{\boxed{}}$

Part 2 ▸ **Plot the line for each equation.**

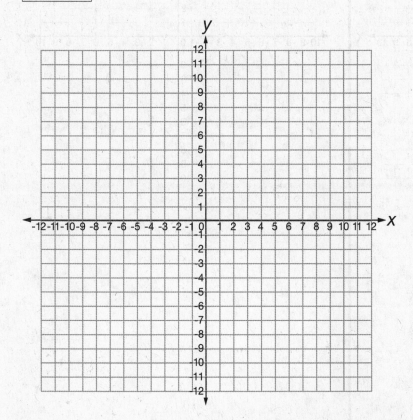

A $y = \dfrac{1}{4}x - 2$

B $y = -2x$

C $y = -\dfrac{4}{5}x$

D $y = \dfrac{8}{3}x + 4$

Lesson 87

Part 1 ▶ Plot the line for each equation.

A $y = -\dfrac{3}{4}x + 3$

B $y = 4x - 4$

C $y = \dfrac{2}{5}x$

D $y = -\dfrac{6}{5}x + 1$

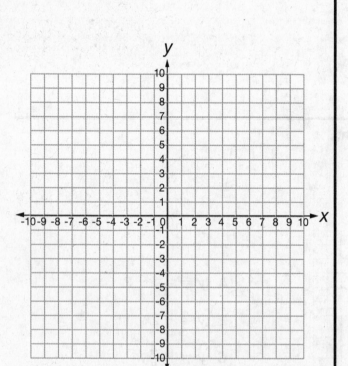

Lesson 88

Part 1 ▶ Plot the line for each equation.

A $y = \dfrac{1}{2}x + 4$

B $y = -3x$

C $y = -\dfrac{5}{4}x - 3$

D $y = x + 2$

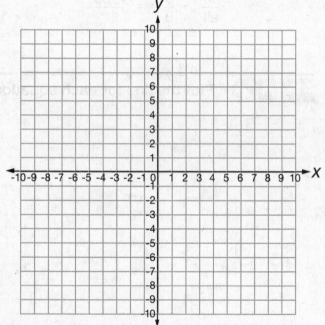

Lesson 89

Independent Work

Part 1 Plot the line for each equation.

A $y = -3x + 5$

B $y = 2x - 1$

C $y = -\dfrac{2}{3}x$

D $y = x - 1$

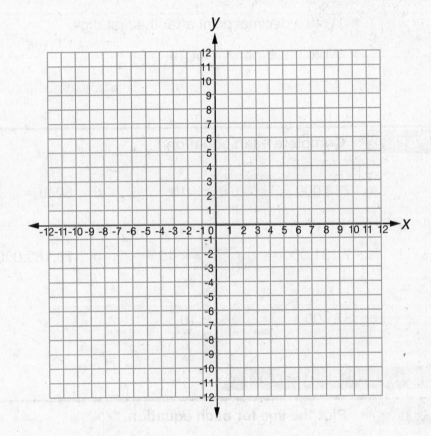

Lesson 92

$50,300 = $ ▮▮▮▮ $\times 10^{■}$

♦ Copy the digits before the final zeros. $5\,03 \times 10^{■}$

♦ Write a decimal point after the first digit. $5.03 \times 10^{■}$

♦ Write the exponent for 10. 5.03×10^4

Part 2 > Complete each equation.

a. $752,000 = $ ▭ $\times 10^{□}$ d. $8040 = $ ▭ $\times 10^{□}$

b. $7,131,000 = $ ▭ $\times 10^{□}$ e. $70,180,000 = $ ▭ $\times 10^{□}$

c. $37,000 = $ ▭ $\times 10^{□}$

> Independent Work <

Part 3 > Plot the line for each equation.

A $y = -\dfrac{3}{4}x - 1$

B $y = x - 4$

C $y = -\dfrac{5}{4}x + 6$

D $y = \dfrac{4}{3}x$

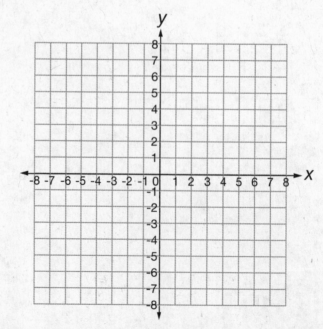

Lesson 93

Part 1 Complete each equation.

a. $20{,}900 = $ ☐ $\times 10^{\blacksquare}$

c. $862{,}030{,}000 = $ ☐ $\times 10^{\blacksquare}$

b. $410 = $ ☐ $\times 10^{\blacksquare}$

d. $500{,}700 = $ ☐ $\times 10^{\blacksquare}$

Lesson 94

Part 1 Plot the line for each equation.

A $y = 2x - 2$

B $y = x$

C $y = -\dfrac{1}{3}x - 4$

D $y = \dfrac{3}{2}x + 5$

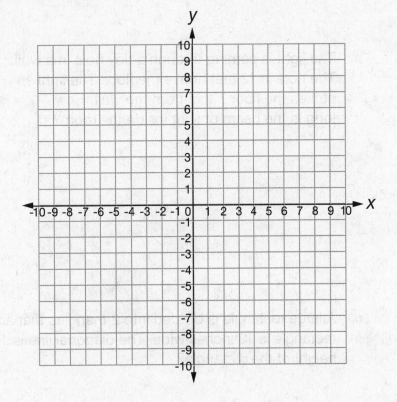

Lesson 95

Part 1 Complete the diagram for each item.
Answer each question.

a. Mary lives 9 miles north of a train station.
   Rick lives 4 miles west of that station.
   How far apart do Mary and Rick live?

Each unit represents 1 mile.

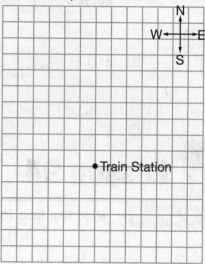

b. The light is coming through a tiny hole in a wall.
   The hole is 12 feet above the floor. The light
   strikes the floor 7 feet from the wall. How
   long is the beam of light inside the room?

Each unit represents 1 foot.

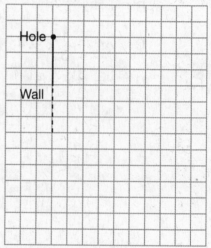

c. A large rectangle is divided into 2 triangles that are the same size. The
   rectangle is 13 inches wide. The diagonal line is 17 inches long. What's the
   height of the rectangle?

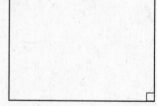

Lesson 96

Part 1 Complete the diagram for each item.
Answer each question.

a. The diagram shows a rectangular cloth that will be folded along the dotted line. The fold is 40 cm long. If the cloth is 20 cm wide, what is the length of the longer side of the cloth?

b. Rob set off from the lodge and skied 6 miles due south. William set off and traveled 8 miles due east. How far apart are the skiers?

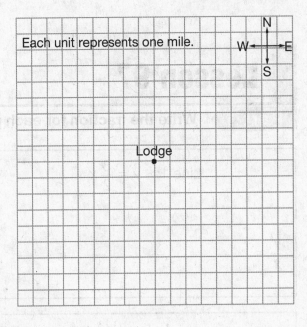

Each unit represents one mile.

N
W ← → E
S

Lodge

c. A window cleaner leans a 12-foot ladder against the wall of a house. The base of the ladder is 3.5 feet from the wall. How far up the wall does the ladder reach?

wall

ground

Plot each line on the coordinate system.

A $y = \dfrac{2}{5}x - 4$

B $y = -\dfrac{3}{2}x + 3$

C $y = 4x - 7$

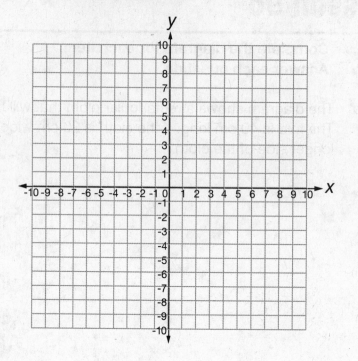

Lesson 97

Part 1 **Write the fraction for each point on lines 1 and 2.**

Line 1: $y = \dfrac{3}{2}x$

$$\dfrac{y}{x} = \underline{\quad} = \underline{\quad} = \underline{\quad} = \underline{\quad}$$

Line 2: $y = \dfrac{1}{4}x$

$$\dfrac{y}{x} = \underline{\quad} = \underline{\quad} = \underline{\quad}$$

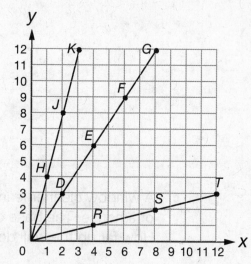

Part 2 ▸ Complete the diagram for each item. Answer each question.

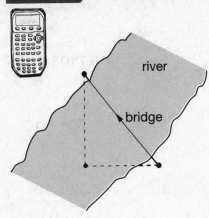

a. The diagram shows a bridge crossing a river. A person walking across the bridge in the direction shown ends up 42 meters north and 35 meters west from where the person started. How long is the bridge?

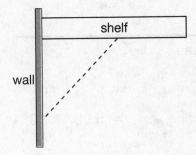

b. A metal support 7 inches long holds up a shelf. The support meets the wall 5 inches below the shelf. At what distance from the wall will the support meet the shelf?

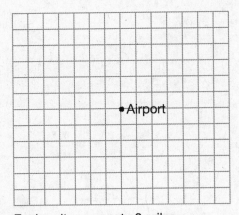

Each unit represents 3 miles.

c. 2 airplanes took off from the airport at the same time. Airplane V went due north. Airplane T went due west. When airplane T had traveled 18 miles from the airport, airplane V had traveled 12 miles from the airport. How far apart were the 2 airplanes at that time?

Complete each equation to show the value equal to the scientific notation.

a. [____] $= 9.0 \times 10^4$ d. [____] $= 5.7108 \times 10^2$

b. [____] $= 9.043 \times 10^2$ e. [____] $= 2.708 \times 10^3$

c. [____] $= 8.6 \times 10^8$ f. [____] $= 4.12 \times 10^6$

Lesson 98

Part 1 Complete each equation to show a number and the scientific notation.

a. [____] $= 2.805 \times 10^3$ d. [____] $= 5.007 \times 10^2$

b. [____] $= 4.29 \times 10^4$ e. 52.401 $=$ [____]

c. 120,000 $=$ [____]   f. 2003 $=$ [____]

Write the fraction for each point.

	A	B	C	D	E
$y = 2x$	$\dfrac{y}{x} = $ ____	= ____	= ____	= ____	= ____

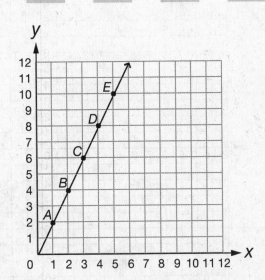

Write each fraction. Circle the fractions for points that would be on the line. Plot and label those points and draw the line.

Equation	(2, 6)	(4, 10)	(4, 12)	(3, 9)	(6, 2)	(5, 15)
$y = 3x$	____ = ____	= ____	= ____	= ____	= ____	= ____

Plot each line.

A $y = \frac{3}{7}x + 5$

B $y = \frac{5}{4}x$

C $y = -3 + \frac{1}{3}x$

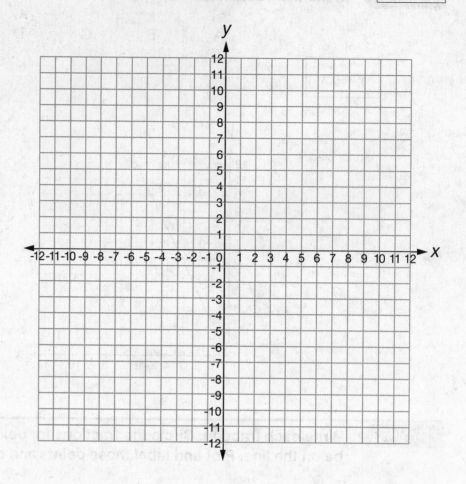

Table 2

	Game #	1	2	3	4	5	Totals
Tigers' star player	free throws attempted Ta	6	4	7	5	6	
	free throws made Tm	5	4	7	4	4	
Cubs' star player	free throws attempted Ca	7	4	7	6	6	
	free throws made Cm	5	3	6	6	4	

◆ This table shows the number of free throws attempted and free throws made by the star players in a 5-game play-off for two teams, the Tigers and the Cubs.

a. The Cubs' star takes 40 free-throw attempts. Based on his performance in the 5 games in Table 2, what's the best prediction of how many free throws he will make?

b. The Tigers' star attempts 35 free throws. Based on his performance in the 5 games in Table 2, what's the best prediction of how many free throws he will make?

c. Over a season, the Tigers' star made 84 free throws. Based on the table data, how many free-throw attempts would you estimate he took during the season?

d. In a season, the Cubs' star made 72 free throws. Based on the table data, how many free throws would you estimate he attempted during the season?

e. If both stars make the same total number of free throws and the Tigers' star has 42 free-throw attempts, how many free throws would you expect the Cubs' star to attempt?

Lesson 99

Part 1 Write each fraction. Circle the fractions for points that would
be on the line. Plot and label those points and draw the line.

Equation   (4, 8)   (9, 12)   (3, 4)   (6, 8)   (4, 3)   (12, 16)

$$y = \frac{4}{3}x$$

——— = ——— = ——— = ——— = ——— = ——— = ———

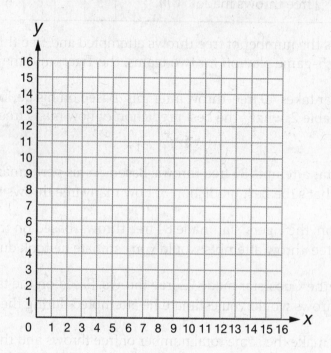

◆ Some numbers have only 1 digit before zeros.

$$50,000 = 5.0 \times 10^4$$
$$50,001 = 5.0001 \times 10^4$$

◆ **Complete each equation.**

a. [_____] $= 5.23 \times 10^7$ d. [_____] $= 8.275 \times 10^2$

b. $1,040,000 =$ [_____]   e. [_____] $= 2.0 \times 10^4$

c. $300,000 =$ [_____]   f. $800.1 =$ [_____]

Independent Work

Part 3 ▷ **Plot each line.**

A $y = -3x - 7$

B $y = \dfrac{2}{3}x + 2$

C $y = -x - 1$

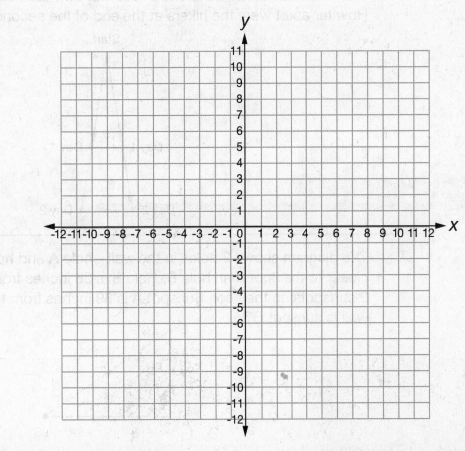

Lesson 100

Part 1 Complete each diagram with 3 numbers and a question mark.
Answer the question the problem asks.

a. The diagram shows 2 poles and their shadows. The length of shadow A is
10 feet. Pole B is 30 feet taller than pole A. Pole A is 15 feet tall. What is the
length of shadow B?

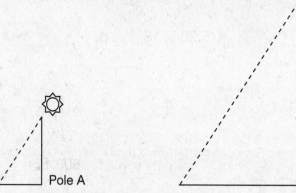

b. The diagram shows the route that Al took and the one Geoff took. Each route
lasted 2 days. On the first day, the hikers went 15 miles. On the second day,
they went 10 miles. At the end of the first day, the hikers were 5 miles apart.
How far apart were the hikers at the end of the second day?

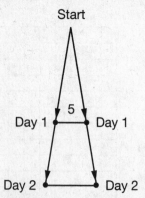

c. The diagram shows 2 holes in the wall—hole A and hole B. Hole A is 10 inches
closer to the floor than hole B. Hole B is 36 inches from the floor. There are
2 sunspots on the floor. Sunspot A is 39 inches from the wall. How far from the
wall is sunspot B?

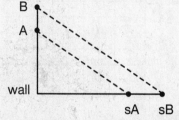

Lesson 104

Part 1 Write the simplified expression below each item.

a. $7^2 + 3r^4 + 2r^4 + r - 7^2$

d. $\frac{3}{5}f^2 - \frac{3}{5}f + \frac{4}{5}f^2$

b. $6m^2 + 3^2 + 5^2 - 5m^2$

e. $2r^7 + r^2 - 2r^2 - 8r^7 + r$

c. $p^5 - 3p^4 - 8p^4$

f. $-3t + r - r^8 + 5t + 4r^8$

For each equation, plot 2 points. Draw the line for B.
Check where the line crosses the *y* axis.

Line A: $y = 2x + 1$

◆ Plot the point for $x = -2$.

◆ Plot the point for $y = 7$.

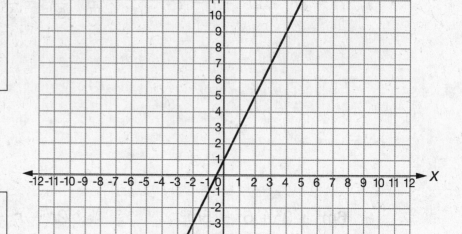

Line B: $y = -x - 5$

◆ Plot the point for $y = -7$.

◆ Plot the point for $x = -8$.

◆ The numbers we'll work with are greater than zero but less than 1.

$$.000064 = \blacksquare \times 10^{\blacksquare}$$

◆ Copy the digits that come **after the zeros.**

◆ Write the **decimal point** after the first digit. $= 6.4 \times 10^{\blacksquare}$

◆ If the original number is **less than 1,** $= 6.4 \times 10^{-5}$
the exponent is **negative.**

Part 4 ➤ Write the scientific notation for each number.

a. 0.00852   = _____

b. .000049   = _____

c. .0000003   = _____

d. .0001008   = _____

e. 0.0706   = _____

> Part 5 > **Plot each line.**

Line A: $y = -\dfrac{3}{2}x + 4$

Line B: $y = 2x - 3$

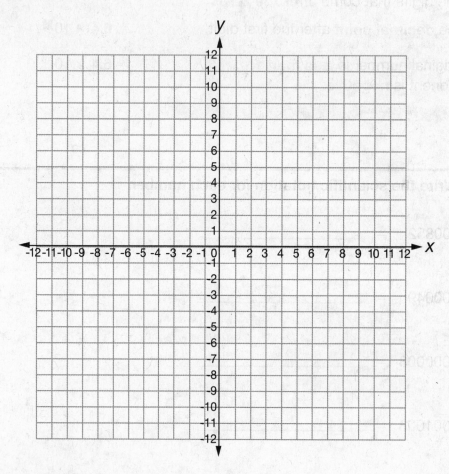

Lesson 105

> Part 1 > **Write the simplified expression below each item.**

    a. $9m^{-2} - 4m^2 + 3m^2 - 5m^{-2}$

    c. $j^{-4} + 4j^5 - 8j^5 + 5j^{-4} - 4^2$

    b. $r^3 - r^2 + 2m^2 + 3r^3 + 2^2$

    d. $-3b + b^3 - 4b^3 + b^2 + 3^3 - 3^2$

For each equation, plot 2 points and draw the line. Check where the line crosses the y axis.

Line A: $y = -\dfrac{1}{2}x + 3$

For one point, $x = -2$.

For another point, $y = 6$.

Line B: $y = 3x - 5$

For one point, $x = 1$.

For another point, $y = 4$.

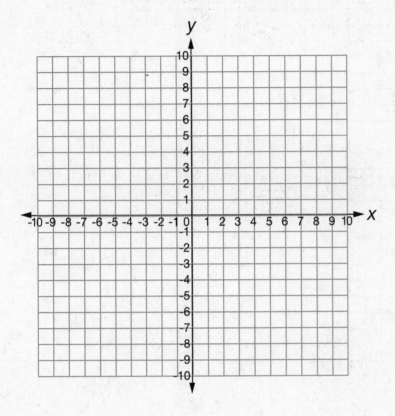

Lesson 106

Part 1 For each equation, plot 2 points and draw the line.
Check where the line crosses the y axis.

Line A: $y = 2x + 1$

For one point, $y = -1$.

For another point, $x = -2$.

Line B: $y = \dfrac{2}{3}x - 8$

For one point, $x = -3$.

For another point, $y = 0$.

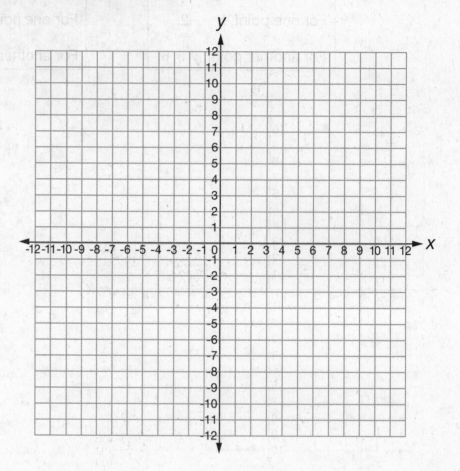

Part 2 ▶ Plot the lines on the graph below.

Line A: $y = \frac{1}{2}x + 2$

Line B: $y = -\frac{1}{4}x - 1$

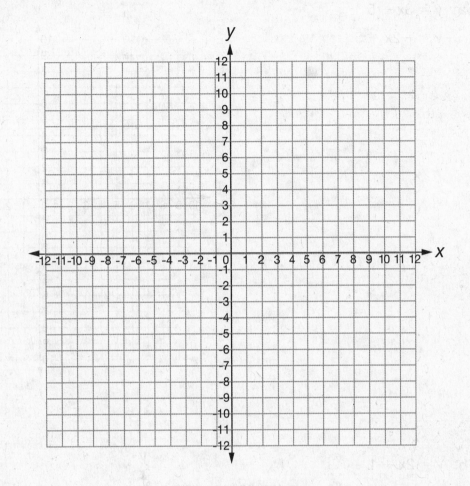

Lesson 107

Part 1 ▶ Write the probabilities for each description.

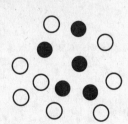

a. white ball =

c. white ball =

e. white ball =

b. black ball =

d. black ball =

f. black ball =

Part 2 Solve each pair of equations for *x* and *y*. Plot the point for the solution. Then draw the line for each equation.

a. $y = 3x - 5$

$y = -2x + 5$

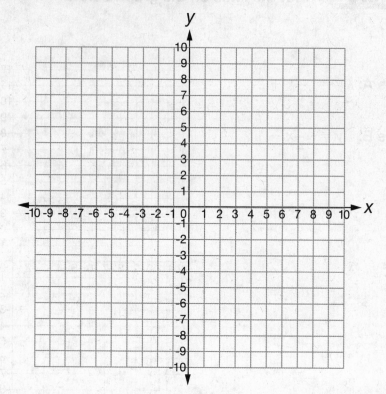

b. $y = 2x + 1$

$y = x + 4$

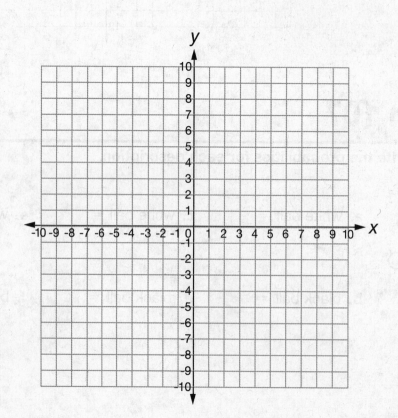

- ◆ Copy the digits shown for the scientific notation.

$$\mathbf{3\ 0\ 8} = 3.08 \times 10^{-5}$$

- ◆ Start after the first digit and count places to the **left.**

$$3\ 0\ 8 =$$

- ◆ Write a new decimal point and a zero for each empty space.

$$.0\ 0\ 0\ 0\ 3\ 0\ 8 =$$

◆ **Write the number for each scientific notation.**

a. $\boxed{712} = 7.12 \times 10^{-6}$ d. $\boxed{} = 1.069 \times 10^{-5}$

b. $\boxed{} = 9.05 \times 10^{-4}$ e. $\boxed{} = 8.3 \times 10^{-7}$

c. $\boxed{} = 4.567 \times 10^{-2}$ f. $\boxed{} = 1.15 \times 10^{-3}$

Lesson 108

Complete each equation to show a number and the equivalent scientific notation.

a. [   ] $= 3.1 \times 10^{-6}$ d. [   ] $= 8.06 \times 10^{-4}$

b. [   ] $= 0.00407$ e. [   ] $= 1.50 \times 10^{5}$

c. $90{,}000{,}000{,}000 =$ [   ]   f. $.03021 =$ [   ]

Part 2 Solve each pair of equations for x and y. Plot the point for the solution. Then draw the line for each equation.

a. $y = -x + 8$

   $y = 2x - 7$

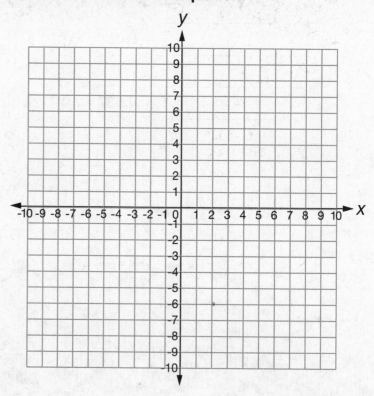

b. $y = 5x + 3$

   $y = 2x$

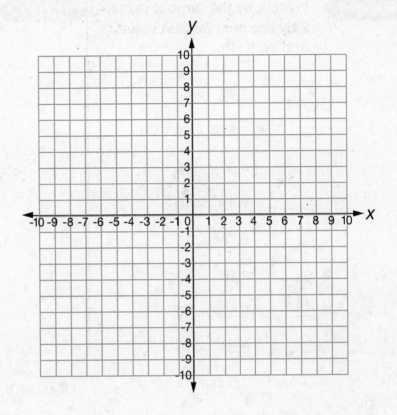

c. $y = x - 4$

   $y = 3x - 4$

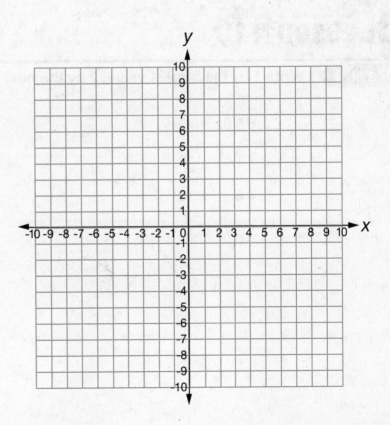

Lesson 109

Write 1 by the largest value, 2 by the next largest value, and so forth.

——— 5.72×10^{-12}

——— 8.524×10^{12}

——— 4.85×10^{-12}

——— 7.4×10^{15}

——— 6.78×10^{17}

——— 2.16×10^{-11}

——— 6.52×10^{17}

——— 6.69×10^{-9}

Write 1 by the largest value, 2 by the next largest value, and so forth.

——— 4.03×10^{6}

——— 2.71×10^{8}

——— 1.18×10^{-3}

——— 3.09×10^{-4}

——— 5.3×10^{6}

——— 9.76×10^{-8}

——— 8×10^{4}

——— 3.5×10^{-4}

Lesson 110

Write 1 by the largest value, 2 by the next largest value, and so forth.

——— 5.01×10^{-6}

——— 1.8×10^{-3}

——— 6×10^{-7}

——— 9.78×10^{2}

——— 3.26×10^{3}

——— 4.73×10^{12}

——— 8.3×10^{-6}

——— 4.621×10^{3}

Lesson 111

Part 1 Use the figure below to answer each question.

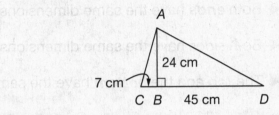

a. What's the length of $\overline{AC}$?

b. What's the length of $\overline{AD}$?

c. What's the **perimeter** of $\triangle ACD$?

d. What's the **area** of $\triangle ACD$?

Lesson 113

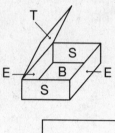

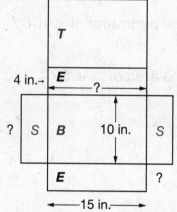

◀ Both **ends** have the same dimensions.

◀ Both *sides* have the same dimensions.

◀ The *top* and the **bottom** have the same dimensions.

◆ Figure out the area of one side, one end, and either the top or bottom.

◆ Add those areas.

◆ Then multiply by 2 to find the total surface area.

Write the length of the parts that have question marks.
Then figure out the surface area of the box.

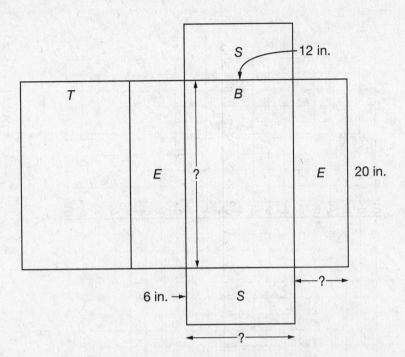

Part 1

a. Show the area for the range x > −1 and y < −7.

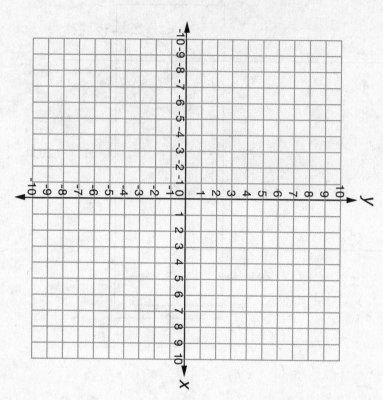

b. Plot and label each point that is within the range x > −1 and y < −7.

Point A (−9, 3)

Point B (7, −8)

Point C (−2, −10)

Point D (2, 8)

Point E (5, −10)

Point F (−2, −6)

Point G (0, −9)

Point H (−4, −11)

Point I (3, 0)

Point J (0, 7)

Lesson 115

For each problem, first convert a larger unit into a smaller unit. Then work the problem.

a. If it takes a crew 2 weeks to lay 2.8 miles of pavement, how much pavement can the crew lay in 20 days?

b. If 5 cups of milk make 3 tarts, how many tarts can be made with 2 pints of milk?

c. The enlargement of the photo is to be 3.5 feet wide and 54 inches high. If the original is 14 inches wide, how high is it?

d. If 7 marbles weigh 8 ounces, how many marbles weigh 2 pounds?

Work each item.

a. Show the area for all points that meet these conditions: **y < 2 and x < 5.**

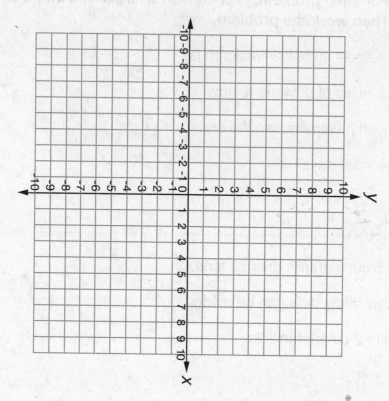

b. Plot and label each point that is within the range **y < 2 and x < 5.**

Point A   (6, 1)

Point B   (3, 3)

Point C   (−1, 0)

Point D   (−5, 6)

Point E   (7, 2)

Point F   (4, 1)

Point G   (5, 3)

Point H   (−2, −3)

Point I   (3, 6)

Point J   (0, 4)

Lesson 116

Part 1 **Work each item.**

a. A recipe calls for 5 eggs and 2 pints of milk. Jane wants to use 1 gallon of milk. How many eggs will she have to use?

b. There are 14 pounds of magazines in every 4 bundles. How much do 30 bundles weigh?

c. It takes a walking team 12 minutes to travel $\frac{1}{4}$ mile. At that rate, how many yards does the team travel in 9 minutes?

d. The leaky faucet drips 1 pint of water every 14 days. How long does it take the faucet to leak 28 fluid ounces of water?

e. Each shelf is 25 feet long. How long are 8 shelves?

Part 2   **Work each item.**

a. Show the area for all points that meet the conditions $y < 1$, $x < 3$.

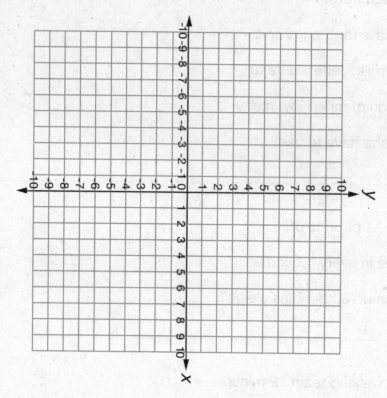

b. Plot and label each point that is within the range $y < 1$, $x < 3$.

Point A: (4, 2)

Point B: (0, 1)

Point C: (−1, 0)

Point D: (1, 0)

Point E: (5, 2)

Point F: (0, 6)

Point G: (3, 4)

Point H: (4, 1)

Point I: (−3, −1)

Point J: (−2, −2)

Lesson 117

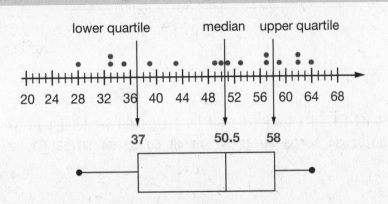

1. Mark the median, the lower quartile, and the upper quartile.

2. Plot a point for the lowest score and the highest score.

3. Draw a box that extends from the lower quartile to the upper quartile.

4. Make a line inside the box for the median score.

5. Draw a line from each side of the box to the lowest score and the highest score.

Answer the questions about this population.

1. a. 50.5 is the number for the _____ .

   b. 37 is the number for the _____ .

   c. 58 is the number for the _____ .

2. a. The lowest score is _____ .

   b. The highest score is _____ .

3. The box extends from the lower _____ to the

   upper _____ .

4. The whisker lines extend from the box to the lowest _____ and the

   highest _____ .

Make a box and whiskers diagram for each population.

a. Here's a set of scores listed from lowest to highest.

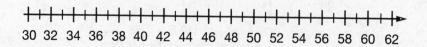

31, 35, 38, 40, 46, 48, 48, 50, 51, 52, 56, 60, 60

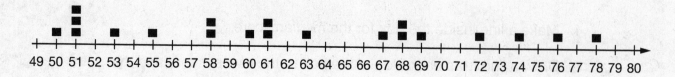

30 32 34 36 38 40 42 44 46 48 50 52 54 56 58 60 62

b. This graph shows a population of 20 scores.

49 50 51 52 53 54 55 56 57 58 59 60 61 62 63 64 65 66 67 68 69 70 71 72 73 74 75 76 77 78 79 80

Part 3 **Work each item.**

a. Show the area for all points that meet the conditions $y < 5$, $x < 0$.

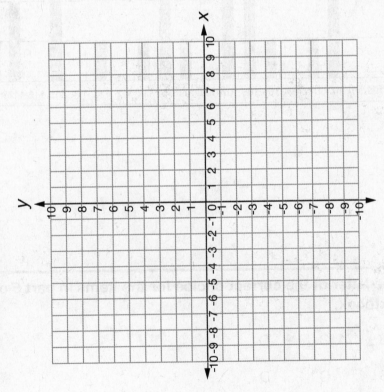

b. Plot and label each point that is within the range $y < 5$, $x < 0$.

Point A: $(-1, 4)$

Point B: $(-2, 2)$

Point C: $(1, 6)$

Point D: $(3, 5)$

Point E: $(-3, 3)$

Point F: $(-5, 0)$

Point G: $(-4, 1)$

Point H: $(0, 6)$

Point I: $(2, 9)$

Point J: $(-6, -3)$

Lesson 118

Part 1 **Make a box and whiskers diagram for each population.**

a.

Scores

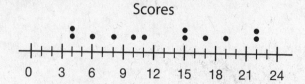

b. The bar graph below shows the distribution of weights for the 21 students in Mr. McKinley's class.

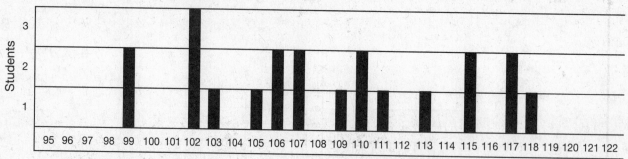

Part 2 **Mark the letter of the correct choice for the items in part 5 of your textbook.**

1. A◯ B◯ C◯ D◯

2. A◯ B◯ C◯ D◯

3. A◯ B◯ C◯ D◯

4. A◯ B◯ C◯ D◯

5. A◯ B◯ C◯ D◯

Part 3 ▸ **Work each item.**

a. Show the area for all the points that meet these conditions: $y < 4$, $x < 0$.

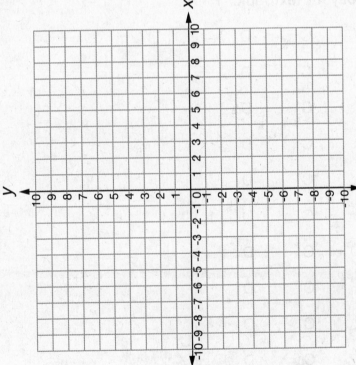

b. Plot and label each point that is within the range $y < 4$, $x < 0$.

Point A: (3, −1)

Point B: (−3, 1)

Point C: (0, −2)

Point D: (3, 2)

Point E: (1, −3)

Point F: (−9, 3)

Lesson 119

Part 1 Mark the letter of the correct choice for the items shown in part 2 of your textbook.

7. A○ B○ C○ D○

8. A○ B○ C○ D○

9. A○ B○ C○ D○

10. A○ B○ C○ D○

11. A○ B○ C○ D○

12. A○ B○ C○ D○

13. A○ B○ C○ D○

Part 2 Mark the letter of the correct choice for the items shown in part 4 of your textbook.

25. A○ B○ C○ D○

26. A○ B○ C○ D○

27. A○ B○ C○ D○

28. A○ B○ C○ D○

29. A○ B○ C○ D○

30. A○ B○ C○ D○

31. A○ B○ C○ D○

32. A○ B○ C○ D○

33. A○ B○ C○ D○

34. A○ B○ C○ D○

35. A○ B○ C○ D○

36. A○ B○ C○ D○

Mark the letter of the correct choice for the items shown in part 6 of your textbook.

51. A○   B○   C○   D○

52. A○   B○   C○   D○

53. A○   B○   C○   D○

54. A○   B○   C○   D○

55. A○   B○   C○   D○

56. A○   B○   C○   D○

57. A○   B○   C○   D○

58. A○   B○   C○   D○

Independent Work

Part 4   Make a box and whiskers diagram for each population.

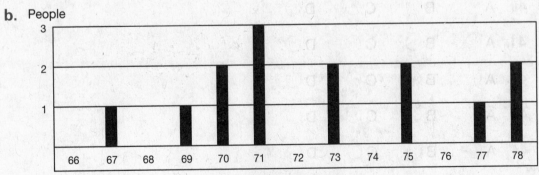

a.

0   3   6   9   12

b. People

Height in Inches

Lesson 120

Part 1 ▶ Mark the letter of the correct choice for the items shown in part 2 of your textbook.

14. A◯   B◯   C◯   D◯
15. A◯   B◯   C◯   D◯
16. A◯   B◯   C◯   D◯
17. A◯   B◯   C◯   D◯
18. A◯   B◯   C◯   D◯
19. A◯   B◯   C◯   D◯
20. A◯   B◯   C◯   D◯

Part 2 ▶ Mark the letter of the correct choice for the items shown in part 4 of your textbook.

35. A◯   B◯   C◯   D◯
36. A◯   B◯   C◯   D◯
37. A◯   B◯   C◯   D◯
38. A◯   B◯   C◯   D◯
39. A◯   B◯   C◯   D◯
40. A◯   B◯   C◯   D◯
41. A◯   B◯   C◯   D◯
42. A◯   B◯   C◯   D◯
43. A◯   B◯   C◯   D◯
44. A◯   B◯   C◯   D◯
45. A◯   B◯   C◯   D◯

Mark the letter of the correct choice for the items shown in part 6 of your textbook.

59. A◯ B◯ C◯ D◯

60. A◯ B◯ C◯ D◯

61. A◯ B◯ C◯ D◯

62. A◯ B◯ C◯ D◯

63. A◯ B◯ C◯ D◯

64. A◯ B◯ C◯ D◯

65. A◯ B◯ C◯ D◯

66. A◯ B◯ C◯ D◯

Independent Work

Part 4 Make a box and whiskers diagram for each population.

a.

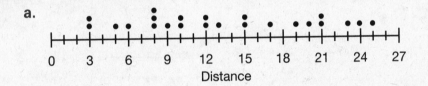

b.

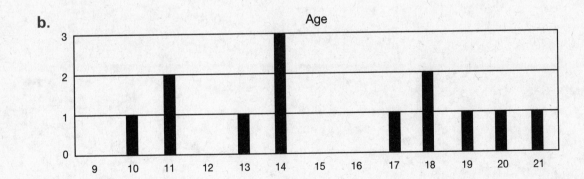

Mastery Test 1B

Part 1 Round each value to the nearest whole number, to the nearest tenth, and to the nearest hundredth.

	whole number	tenths	hundredths
49.1052			
0.251			
9.6958			

Part 2 Write a decimal point to show the correct number of decimal places in the answer. You may have to write one or more zeros.

a.
$$\begin{array}{r} 5.2 \\ \times\,.0\,3 \\ \hline 1\,5\,6 \end{array}$$

b.
$$\begin{array}{r} 1\,4.2 \\ \times\,\ \ .7 \\ \hline 9\,9\,4 \end{array}$$

c.
$$\begin{array}{r} 2.4 \\ \times\,.3\,0\,9 \\ \hline 7\,4\,1\,6 \end{array}$$

d.
$$\begin{array}{r} 7.0\,3 \\ \times\,.0\,0\,9\,1 \\ \hline 6\,3\,9\,7\,3 \end{array}$$

Part 3 Complete each row to show the equivalent decimal, fraction, and percent.

Decimal	Fraction	Percent
3.15		
	$\frac{7}{100}$	
		40%
		3%
	$\frac{200}{100}$	

This test is continued in the textbook on page T2.

Mastery Test 3

Part 1 — Make a point on the coordinate system for each description. Label each point.

Point A: $x = 6$, $y = 4$
Point B: $x = 9$, $y = 3$
Point C: $x = 5$, $y = 0$
Point D: $x = 0$, $y = 7$

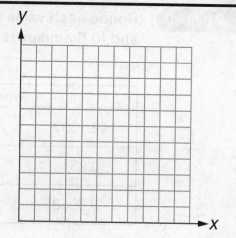

Part 2 — Simplify each problem. Then work it.

a. $\dfrac{5}{6}\left(\dfrac{12}{3}\right) =$

c. $\dfrac{4r}{9}\left(\dfrac{3}{68}\right) =$

e. $\dfrac{25}{35} - \dfrac{1}{7} =$

b. $\dfrac{3}{4} + \dfrac{28}{16} =$

d. $\dfrac{5}{6} + \dfrac{24}{12} =$

f. $10y\left(\dfrac{8}{40y}\right) =$

Part 3 — Figure out what each letter equals.

a. $2p = 14 - 8$

c. $4 + 3m = 25$

e. $17 + 11 = 7r$

b. $2 = 4d - 10$

d. $\dfrac{1}{3}m - 4 = 1$

f. $9 = \dfrac{2}{3}p + 3$

This test is continued in the textbook on page T6.

Mastery Test 4

Part 1 Complete the function table. Plot the points and draw the line.

Function		
x	$-3 =$	y
A	3	
B	9	
C	5	
D	13	

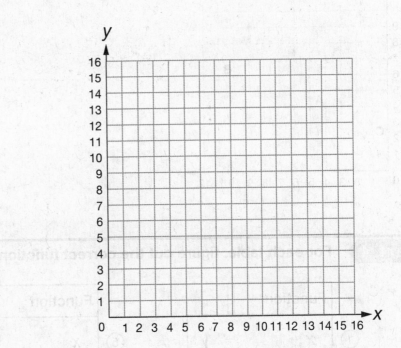

Part 2 Complete each equation.

a. $-10 + 30 - 20 + 40 =$

c. $-6 + 2 - 8 + 3 =$

b. $-51 + 35 - 0 =$

Write the description for points A–C.
Plot points D–F on the coordinate system.

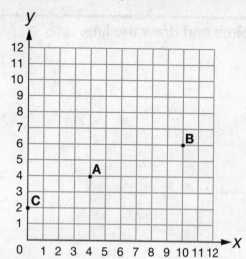

A

B

C

D (12, 0)

E (3, 2)

F (0, 7)

For each table, figure out the correct function. Complete the table.

Function		
① X	=	y
② X	=	y
9		3
3		1
③ 6		
④ 24		

Function		
⑤ X	=	y
⑥ X	=	y
4		8
2		6
⑦ 3		
⑧ 0		

This test is continued in the textbook on page T8.

Mastery Test 5

Part 1 > Write the base and exponent for each fraction.

a. $\dfrac{p \times p \times p \times p}{p \times p \times p \times p \times p \times p \times p} =$ ▢

b. $\dfrac{4 \times 4 \times 4}{4 \times 4 \times 4} =$ ▢

c. $\dfrac{11 \times 11}{11 \times 11 \times 11 \times 11} =$ ▢

d. $\dfrac{6 \times 6 \times 6 \times 6}{6} =$ ▢

Part 2 > Figure out the function. Complete the table.

Function		
x $\left(\dfrac{y}{x}\right)$	=	y
① x $\left(\dfrac{\ \ \ }{\ \ \ }\right)$	=	y
3		2
②		18
③ 12		
④		6

Part 3 > Complete each boxed equation.

a. $(7 \times 7 \times 7) \times (7) \times (7 \times 7 \times 7 \times 7)$

$7^{▢} =$ ▢ $\times$ ▢ $\times$ ▢

b. $(m \times m) \times (m \times m \times m \times m \times m)$

$m^{▢} =$ ▢ $\times$ ▢

This test is continued in the textbook on page T10.

Mastery Test 6

Part 1 ▶ Rewrite each fraction. First show a base with a positive exponent. Then show a base with a negative exponent.

a. $\dfrac{6 \times 6 \times 6}{6 \times 6 \times 6 \times 6}$ = ⬜ = ⬜

c. $\dfrac{11 \times 11 \times 11 \times 11 \times 11 \times 11}{11 \times 11}$ = ⬜ = ⬜

b. $\dfrac{k \times k \times k \times k \times k}{k \times k \times k}$ = ⬜ = ⬜

d. $\dfrac{m}{m \times m \times m}$ = ⬜ = ⬜

Part 2 ▶ Complete the table. Plot and label each point. Show the coordinates. Then draw the line.

	Function
	$y = \dfrac{1}{2} x$
A	4	
B		6
C		10
D	2	

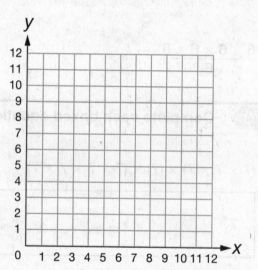

Part 3 ▶ Work each problem.

a. $6(-3) + 2(5) - 4\left(\dfrac{7}{2}\right) = $ ■

⬜ = ⬜

c. $-6\left(-\dfrac{1}{3}\right) + 2 - 8(-5) = $ ■

⬜ = ⬜

b. $-8 - 30(-1) + 6(+4) = $ ■

⬜ = ⬜

⟨ This test is continued in the textbook on page T12. ⟩

Mastery Test 7

Part 1 Plot a point for each line. Then write the complete equation for each line.

A

B

C

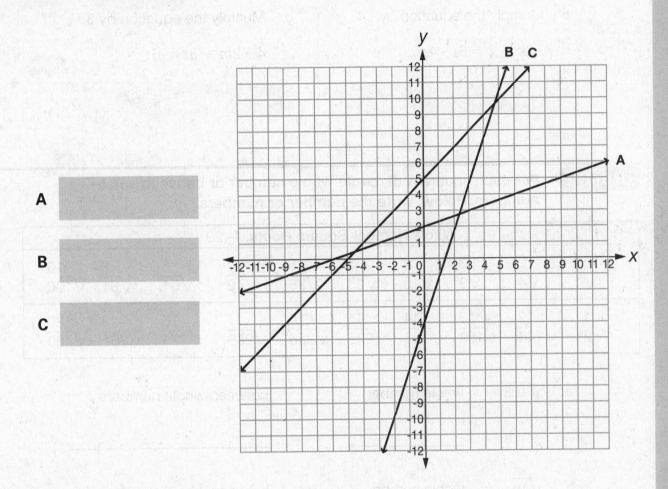

Part 2 Rewrite each expression below with combined exponents.

a. $x^{-4}\, pn^2\, n^{-6}\, p^3$ b. $5^4\, m^9\, m^{-6}\, 5^{-3}\, k$ c. $y^{-2}\, d^9\, y^{-8}\, y^5\, d^2$

This test is continued in the textbook on page T14.

Mastery Test 8

Multiply both sides of the equation. Remove the parentheses.

a. Multiply the equation by -4.

$$-3k + 2 = 6j - 4p$$

b. Multiply the equation by 3.

$$4 - 2m + 3z = 6t$$

Part 2 For each square root, circle **whole number or between whole numbers.** Below, write the number or numbers.

Whole Number Square Roots, 1–20									
1 $\sqrt{1}$	2 $\sqrt{4}$	3 $\sqrt{9}$	4 $\sqrt{16}$	5 $\sqrt{25}$	6 $\sqrt{36}$	7 $\sqrt{49}$	8 $\sqrt{64}$	9 $\sqrt{81}$	10 $\sqrt{100}$
11 $\sqrt{121}$	12 $\sqrt{144}$	13 $\sqrt{169}$	14 $\sqrt{196}$	15 $\sqrt{225}$	16 $\sqrt{256}$	17 $\sqrt{289}$	18 $\sqrt{324}$	19 $\sqrt{361}$	20 $\sqrt{400}$

a. $\sqrt{109}$ whole number   between whole numbers

_____   _____

b. $\sqrt{16}$ whole number   between whole numbers

_____   _____

c. $\sqrt{56}$ whole number   between whole numbers

_____   _____

d. $\sqrt{250}$ whole number   between whole numbers

_____   _____

Label the *x* and *y* axes. Then make a line for each equation.
Label each line.

A $y = -x - 3$

B $y = \dfrac{4}{5}x + 1$

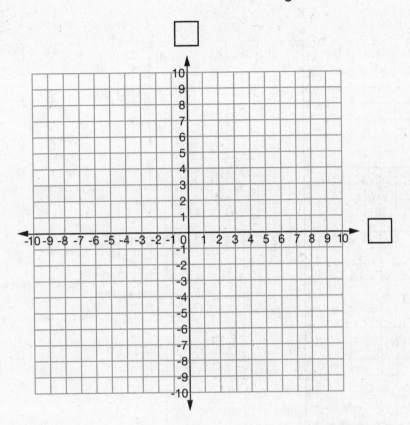

Part 4 Complete each equation.

a. $-5m\,(\boxed{}) = +15m$

c. $-4j\,(\boxed{}) = -36j$

b. $+t\,(\boxed{}) = -14t$

d. $-3q\,(\boxed{}) = 12q$

⟨ **This test is continued in the textbook on page T17.** ⟩

Mastery Test 9

Draw the line for each equation. Label each line.

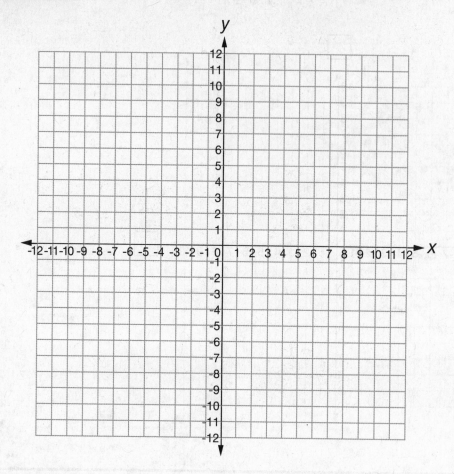

A $y = \dfrac{1}{3}x$

B $y = 2x$

C $y = -x + 3$

Part 2 **Complete each equation.**

a. $.024601 \times 10^{\blacksquare} = 2460.1$

b. $3.14159 \times 10^6 = $

c. $265.35 \times 10^{\blacksquare} = 26{,}535{,}000{,}000$

d. $9.0703 \times 10^3 = $

⟨ **This test is continued in the textbook on page T19.** ⟩

Mastery Test 10

Complete each equation so it shows scientific notation.

a. $3280 = $ _____

b. _____ $= 6.13 \times 10^5$

c. _____ $= 1.0481 \times 10^3$

d. $1,000,000 = $ _____

e. $34.51 = $ _____

Part 2 Write the simplified expression below each item.

a. $3r^5 + r^3 - 4r^3 + 6r^5 - r^2$

b. $m^4 - 4b - t^2 + 2b + 8t^2 + 4^2 - 4$

c. $\frac{2}{5}p - \frac{1}{5}p + hp^4 - 3hp^4 + \frac{3}{5}p$

⟨ This test is continued in the textbook on page T23. ⟩

Mastery Test 11

Part 1 Solve for *x* and *y*, plot 2 points, and draw the line. Check the *y*-intercept.

Line A: $y = \dfrac{1}{2}x + 3$

For one point, $y = 2$.

For another point, $x = 4$.

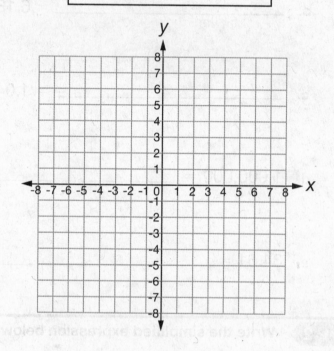

Part 2 Write the equation for line A. Then write the *y*-intercept.

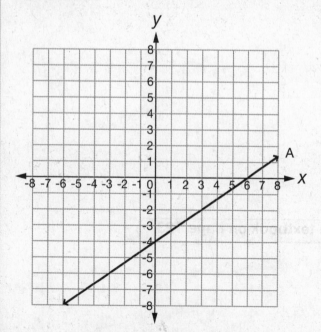

Equation:

y-intercept:

Part 3 ⟩ Solve each pair of equations for *x* and *y*. Plot the point for the solution. Then draw the line for each equation.

a. $y = 2x - 7$

$y = -x + 2$

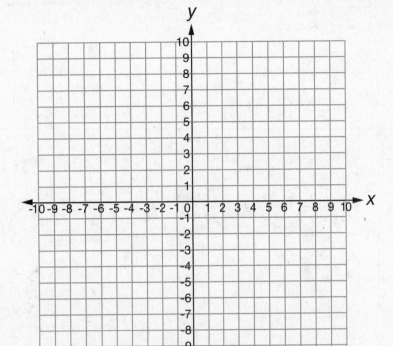

b. $y = -x + 7$

$y = 3x + 3$

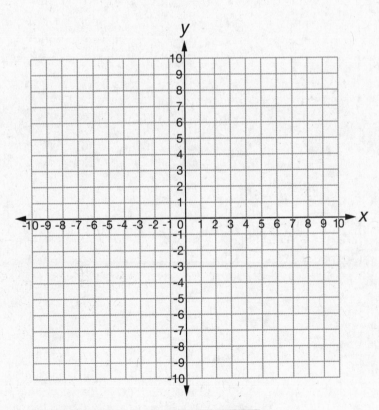

⟨ **This test is continued in the textbook on page T27.** ⟩

Practice Test 1

Mark the letter of the correct choice for the items shown in
Practice Test 1.

1. A○	B○	C○	D○	E○		31. A○	B○	C○	D○	E○
2. A○	B○	C○	D○	E○		32. A○	B○	C○	D○	E○
3. A○	B○	C○	D○	E○		33. A○	B○	C○	D○	E○
4. A○	B○	C○	D○	E○		34. A○	B○	C○	D○	E○
5. A○	B○	C○	D○	E○		35. A○	B○	C○	D○	E○
6. A○	B○	C○	D○	E○		36. A○	B○	C○	D○	E○
7. A○	B○	C○	D○	E○		37. A○	B○	C○	D○	E○
8. A○	B○	C○	D○	E○		38. A○	B○	C○	D○	E○
9. A○	B○	C○	D○	E○		39. A○	B○	C○	D○	E○
10. A○	B○	C○	D○	E○		40. A○	B○	C○	D○	E○
11. A○	B○	C○	D○	E○		41. A○	B○	C○	D○	E○
12. A○	B○	C○	D○	E○		42. A○	B○	C○	D○	E○
13. A○	B○	C○	D○	E○		43. A○	B○	C○	D○	E○
14. A○	B○	C○	D○	E○		44. A○	B○	C○	D○	E○
15. A○	B○	C○	D○	E○		45. A○	B○	C○	D○	E○
16. A○	B○	C○	D○	E○		46. A○	B○	C○	D○	E○
17. A○	B○	C○	D○	E○		47. A○	B○	C○	D○	E○
18. A○	B○	C○	D○	E○		48. A○	B○	C○	D○	E○
19. A○	B○	C○	D○	E○		49. A○	B○	C○	D○	E○
20. A○	B○	C○	D○	E○		50. A○	B○	C○	D○	E○
21. A○	B○	C○	D○	E○		51. A○	B○	C○	D○	E○
22. A○	B○	C○	D○	E○		52. A○	B○	C○	D○	E○
23. A○	B○	C○	D○	E○		53. A○	B○	C○	D○	E○
24. A○	B○	C○	D○	E○		54. A○	B○	C○	D○	E○
25. A○	B○	C○	D○	E○		55. A○	B○	C○	D○	E○
26. A○	B○	C○	D○	E○		56. A○	B○	C○	D○	E○
27. A○	B○	C○	D○	E○		57. A○	B○	C○	D○	E○
28. A○	B○	C○	D○	E○		58. A○	B○	C○	D○	E○
29. A○	B○	C○	D○	E○		59. A○	B○	C○	D○	E○
30. A○	B○	C○	D○	E○		60. A○	B○	C○	D○	E○

Practice Test 2

Mark the letter of the correct choice for the items shown in
Practice Test 2.

1. A○ B○ C○ D○ E○   31. A○ B○ C○ D○ E○
2. A○ B○ C○ D○ E○   32. A○ B○ C○ D○ E○
3. A○ B○ C○ D○ E○   33. A○ B○ C○ D○ E○
4. A○ B○ C○ D○ E○   34. A○ B○ C○ D○ E○
5. A○ B○ C○ D○ E○   35. A○ B○ C○ D○ E○
6. A○ B○ C○ D○ E○   36. A○ B○ C○ D○ E○
7. A○ B○ C○ D○ E○   37. A○ B○ C○ D○ E○
8. A○ B○ C○ D○ E○   38. A○ B○ C○ D○ E○
9. A○ B○ C○ D○ E○   39. A○ B○ C○ D○ E○
10. A○ B○ C○ D○ E○   40. A○ B○ C○ D○ E○
11. A○ B○ C○ D○ E○   41. A○ B○ C○ D○ E○
12. A○ B○ C○ D○ E○   42. A○ B○ C○ D○ E○
13. A○ B○ C○ D○ E○   43. A○ B○ C○ D○ E○
14. A○ B○ C○ D○ E○   44. A○ B○ C○ D○ E○
15. A○ B○ C○ D○ E○   45. A○ B○ C○ D○ E○
16. A○ B○ C○ D○ E○   46. A○ B○ C○ D○ E○
17. A○ B○ C○ D○ E○   47. A○ B○ C○ D○ E○
18. A○ B○ C○ D○ E○   48. A○ B○ C○ D○ E○
19. A○ B○ C○ D○ E○   49. A○ B○ C○ D○ E○
20. A○ B○ C○ D○ E○   50. A○ B○ C○ D○ E○
21. A○ B○ C○ D○ E○   51. A○ B○ C○ D○ E○
22. A○ B○ C○ D○ E○   52. A○ B○ C○ D○ E○
23. A○ B○ C○ D○ E○   53. A○ B○ C○ D○ E○
24. A○ B○ C○ D○ E○   54. A○ B○ C○ D○ E○
25. A○ B○ C○ D○ E○   55. A○ B○ C○ D○ E○
26. A○ B○ C○ D○ E○   56. A○ B○ C○ D○ E○
27. A○ B○ C○ D○ E○   57. A○ B○ C○ D○ E○
28. A○ B○ C○ D○ E○   58. A○ B○ C○ D○ E○
29. A○ B○ C○ D○ E○   59. A○ B○ C○ D○ E○
30. A○ B○ C○ D○ E○   60. A○ B○ C○ D○ E○